HANDBOOK ON BATTERY ENERGY STORAGE SYSTEM

DECEMBER 2018

ASIAN DEVELOPMENT BANK

© 2018 Asian Development Bank
6 ADB Avenue, Mandaluyong City, 1550 Metro Manila, Philippines
Tel +63 2 632 4444; Fax +63 2 636 2444
www.adb.org

Some rights reserved. Published in 2018.
ISBN 978-92-9261-470-6 (print), 978-92-9261-471-3 (electronic)
Publication Stock No. TCS189791-2
DOI: http://dx.doi.org/10.22617/TCS189791-2

Notes:
In this publication, "$" refers to United States dollars.
ADB recognizes "Korea" as the Republic of Korea.
On the cover: ADB Solar Mini Grid Pilot Project in Harkapur, Okhaldhunga, Nepal (Photo by C. Lao Torregosa); and, ADB solar-wind hybrid project site in Pira Kalwal and Wadgal Village, Joharabad, Khushab District, Pakistan (Photo by Nasr ur Rahman)
Corrigenda to ADB publications may be found at http://www.adb.org/publications/corrigenda.

 Printed on recycled paper

CONTENTS

Appendixes

TABLES AND FIGURES

Tables

Figures

EXECUTIVE SUMMARY

This handbook serves as a guide to the applications, technologies, business models, and regulations that should be considered when evaluating the feasibility of a battery energy storage system (BESS) project.

Several applications and use cases, including frequency regulation, renewable integration, peak shaving, microgrids, and black start capability, are explored. For example, the integration of distributed energy resources into traditional unidirectional electric power systems is challenging because of the increased complexity of maintaining system reliability despite the variable and intermittent nature of wind and solar power generation, or keeping customer tariffs affordable while investing in network expansion, advanced metering infrastructure, and other smart grid technologies.

The key to overcoming such challenges is to increase power system flexibility so that the occasional periods of excessive renewable power generation need not be curtailed or so that there is less need for large investments in network expansion that lead to high consumer prices. Storage offers one possible source of flexibility.

Batteries have already proven to be a commercially viable energy storage technology. BESSs are modular systems that can be deployed in standard shipping containers. Until recently, high costs and low round trip efficiencies prevented the mass deployment of battery energy storage systems. However, increased use of lithium-ion batteries in consumer electronics and electric vehicles has led to an expansion in global manufacturing capacity, resulting in a significant cost decrease that is expected to continue over the next few years. The low cost and high efficiency of lithium-ion batteries has been instrumental in a wave of BESS deployments in recent years for both small-scale, behind-the-meter installations and large-scale, grid-level deployments. This handbook breaks down the BESS into its critical components and provides a basis for estimating the costs of future BESS projects.

For example, battery energy storage systems can be used to overcome several challenges related to large-scale grid integration of renewables. First, batteries are technically better suited to frequency regulation than the traditional spinning reserve from power plants. Second, batteries provide a cost-effective alternative to network expansion for reducing curtailment of wind and solar power generation. Similarly, batteries enable consumer peak charge avoidance by supplying off-grid energy during on-grid peak consumption hours. Third, as renewable power generation often does not coincide with electricity demand, surplus power should be either curtailed or exported. Surplus power can instead be stored in batteries for consumption later when renewable power generation is low and electricity demand increases. The financial viability of a BESS project for renewable integration will depend on the cost–benefit analysis of the intended application.

The business case for battery energy storage differs by application and by use case. "Prosumers" (producers–consumers) can calculate the payback period of a home energy storage system from the spread between the cost of producing and storing rooftop solar power and the cost of purchasing electricity from the local utility. Industrial consumers and distribution network owners benefit from a reduction in peak capacity charges and network expansion deferral because of peak shaving and load leveling. The business case for using batteries for frequency regulation depends on revenue forecasts and competition for ancillary services.

Various business models are possible, depending on how ownership and operations responsibility is divided between utility customers or prosumers and the utility or network operator. For example, while the charge and discharge cycles of home energy storage systems are set by the home owners themselves, industrial battery systems could be operated by a demand-side management provider or flexibility aggregator. Similarly, while large-scale batteries used for frequency regulation may be owned by private investors, the operation of such systems is likely to be the responsibility of the transmission system operator as part of the pool of assets that provide spinning reserve.

This handbook lists the major policy and regulatory changes that could help promote energy storage markets and projects. For example, in most countries that operate ancillary service markets, frequency regulation products have historically been designed with the technical limitations of large power stations in mind. However, in 2016, a new ancillary service known as "enhanced frequency response," with a sub-second response time that could be met only with the help of batteries, was launched in Europe. Similarly, in some countries, the provision of frequency regulation is mandatory for developers of large wind farms, to reduce the need for increased spinning reserve from conventional power plants.

As with most projects, it is important to capture the risks and challenges in undertaking a typical battery energy storage project. This handbook outlines the most important risks and challenges from a project execution perspective. It also provides a guide for building a financial model for BESS projects, including popular investment metrics such as the levelized cost of storage.

ABBREVIATIONS

AC   –   alternating current

ACB   –   air circuit breaker

BESS   –   battery energy storage system

BMS   –   battery management system

CAES   –   compressed air energy storage

CB   –   circuit breaker

C&I   –   commercial and industrial

COD   –   commercial operation date

DLC   –   double-layer capacitor

EFR   –   enhanced frequency response

EIS   –   electric insulation switchgear

EMS   –   energy management system

EPC   –   engineering, procurement, and construction

ESCO   –   energy service company

ESS   –   energy storage system

EV   –   electric vehicle

FES   –   flywheel energy storage

GIS   –   gas insulation switchgear

HSCB   –   high-speed circuit breaker

IGBT   –   insulated gate bipolar transistors

IPP   –   independent power producer

kW   –   kilowatt

kWh   –   kilowatt–hour

LA   –   lead–acid

LCOS   –   levelized cost of energy storage

LFP   –   lithium–iron–phosphate

LMO   –   lithium–manganese oxide

LPMS   –   local power management system

LSE   –   load-serving entity

LTO   –   lithium–titanate

MW   –   megawatt

NCA   –   nickel–cobalt–aluminum oxide

PCC   –   point of common coupling

PCS   –   power conversion system

PMS   –   power management system

PV   –   photovoltaic

SCS   –   supervisory control system

SOC   –   state of charge

SOH   –   state of heath

UPS   –   uninterruptible power supply

VRFB   –   vanadium redox flow battery

VRLA   –   valve-regulated lead–acid

W   –   watt

ZBFB   –   zinc-bromine flow battery

1 ENERGY STORAGE TECHNOLOGIES

This chapter provides an overview of commonly used energy storage technologies. It looks into various factors that differentiate storage technologies, such as cost, cycle life, energy density, efficiency, power output, and discharge duration.

One energy storage technology in particular, the battery energy storage system (BESS), is studied in greater detail together with the various components required for grid-scale operation. The advantages and disadvantages of different commercially mature battery chemistries are examined. The chapter ends with a review of best practice for recycling and reuse lithium-ion batteries.

1.1 STORAGE TYPES

Energy storage devices can be categorized as mechanical, electrochemical, chemical, electrical, or thermal devices, depending on the storage technology used (Figure 1.1). Mechanical technology, including pumped hydropower generation, is the oldest technology. However, a limitation of this technology is its need for abundant water resources and a different geographic elevation, as well as the construction of power transmission lines to households that consume electricity. Recently, transmission-line construction cost has surpassed the cost of installing a pumped hydropower generation facility.

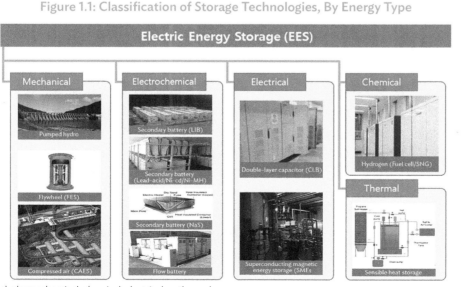

Figure 1.1: Classification of Storage Technologies, By Energy Type

*Mechanical, electrochemical, chemical, electrical, or thermal.
Li-ion = lithium-ion, Na–S = sodium–sulfur, Ni–CD = nickel–cadmium, Ni–MH = nickel–metal hydride, SMES=superconducting magnetic energy storage.
Source: Korea Battery Industry Association 2017 "Energy storage system technology and business model".

In addition to the recent spread of mobile information technology (IT) devices and electric vehicles, the increased mass production of lithium secondary batteries and their lowered costs have boosted demand for energy storage devices using such batteries. Lithium secondary batteries convert electric energy to chemical energy, and vice versa, using electrochemical technologies. Such technologies also include lead storage batteries and sodium–sulfur batteries. Chemical technologies include energy storage technologies such as fuel cells, and mechanical technologies include electric double-layer capacitors.

The performance of energy storage devices can be defined by their output and energy density. Their use can be differentiated by place and duration of use, as defined by the technology adopted. In Figure 1.2, the applications (in the tan-colored boxes) are classified according to output, usage period, and power requirement, and the energy storage devices (in the amber-colored boxes) according to usage period, power generation, and system and/or network operation.

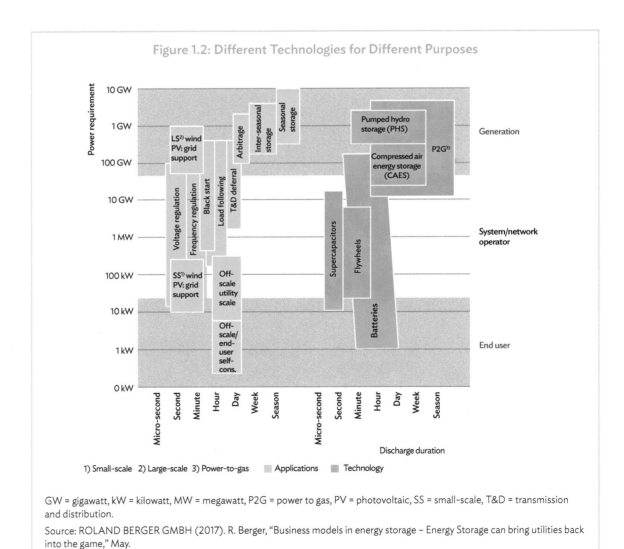

Figure 1.2: Different Technologies for Different Purposes

1) Small-scale   2) Large-scale   3) Power-to-gas   ▧ Applications   ▨ Technology

GW = gigawatt, kW = kilowatt, MW = megawatt, P2G = power to gas, PV = photovoltaic, SS = small-scale, T&D = transmission and distribution.

Source: ROLAND BERGER GMBH (2017). R. Berger, "Business models in energy storage – Energy Storage can bring utilities back into the game," May.

Energy storage devices can be used for uninterruptible power supply (UPS), transmission and distribution (T&D) system support, or large-scale generation, depending on the technology applied and on storage capacity. Among electrochemical, chemical, and physical energy storage devices, the technologies that have received the most attention recently fall within the scope of UPS and T&D system support (Figure 1.3). Representative technologies include reduction–oxidation (redox) flow, sodium–sulfur (Na–S), lead–acid and advanced lead–acid, super-capacitor, lithium, and flywheel batteries. Lithium batteries are in common use today.

Figure 1.3: Comparison of Power Output (in watts) and Energy Consumption (in watt-hours) for Various Energy Storage Technologies

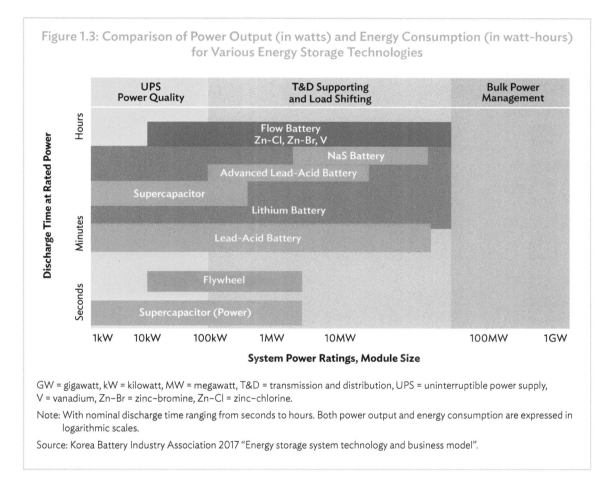

GW = gigawatt, kW = kilowatt, MW = megawatt, T&D = transmission and distribution, UPS = uninterruptible power supply, V = vanadium, Zn–Br = zinc–bromine, Zn–Cl = zinc–chlorine.

Note: With nominal discharge time ranging from seconds to hours. Both power output and energy consumption are expressed in logarithmic scales.

Source: Korea Battery Industry Association 2017 "Energy storage system technology and business model".

Battery technologies for energy storage devices can be differentiated on the basis of energy density, charge and discharge (round trip) efficiency, life span, and eco-friendliness of the devices (Figure 1.4). Energy density is defined as the amount of energy that can be stored in a single system per unit volume or per unit weight. Lithium secondary batteries store 150–250 watt-hours per kilogram (kg) and can store 1.5–2 times more energy than Na–S batteries, two to three times more than redox flow batteries, and about five times more than lead storage batteries.

Charge and discharge efficiency is a performance scale that can be used to assess battery efficiency. Lithium secondary batteries have the highest charge and discharge efficiency, at 95%, while lead storage batteries are at about 60%–70%, and redox flow batteries, at about 70%–75%.

One important performance element of energy storage devices is their life span, and this factor has the biggest impact in reviewing economic efficiency. Another major consideration is eco-friendliness, or the extent to which the devices are environmentally harmless and recyclable.

Figure 1.4: Differentiating Characteristics of Different Battery Technologies

	Energy density (kW/kg)	Round Trip Efficiency (%)	Life Span (years)	Eco-friendliness
Li-ion	1st (150–250)	1st 95	1st (10–15)	Eco-friendliness
NaS	2nd (125–150)	2nd (75–85)	3rd (10–15)	X
Flow	2nd (60–80)	2nd (70–75)	3rd (20–25)	X
Ni-Cd	4th (40–60)	4th (60–80)	4th (5–10)	X
Lead Acid	5th (30–50)	5th (60–70)	5th (3–6)	X

Li-ion = lithium-ion, Na–S = sodium–sulfur, Ni–Cd = nickel–cadmium.

Source: Korea Battery Industry Association 2017 "Energy storage system technology and business model".

Technological changes in batteries are progressing toward higher energy density (Figure 1.5). Next-generation battery technologies—lithium-ion, zinc–air, lithium–sulfur, lithium–air, etc.—are expected to improve on the energy density of lithium secondary (rechargeable) batteries, and be priced below $50 per kilowatt (kW).

Energy storage device applications vary depending on the time needed to connect to the generator, transmitter, and place of use of energy, and on energy use. Black start, a technology for restarting generators after blackouts without relying on the external power grid, is installed in the generating bus and supplies energy within 15–30 minutes. Power supply for maintaining frequency is provided within a quarter-hour to an hour of system operation. Power supply for maintaining voltage level is provided within a shorter operating interval. Grid storage needs are categorized in Figure 1.6 according to network function, power market, and duration of use. Table 1.1 compares the various battery technologies according to discharge time and energy-to-power ratio.

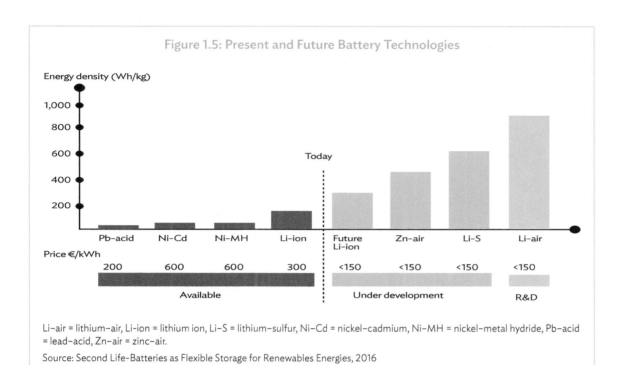

Figure 1.5: Present and Future Battery Technologies

Li–air = lithium–air, Li–ion = lithium ion, Li–S = lithium–sulfur, Ni–Cd = nickel–cadmium, Ni–MH = nickel–metal hydride, Pb–acid = lead–acid, Zn–air = zinc–air.

Source: Second Life-Batteries as Flexible Storage for Renewables Energies, 2016

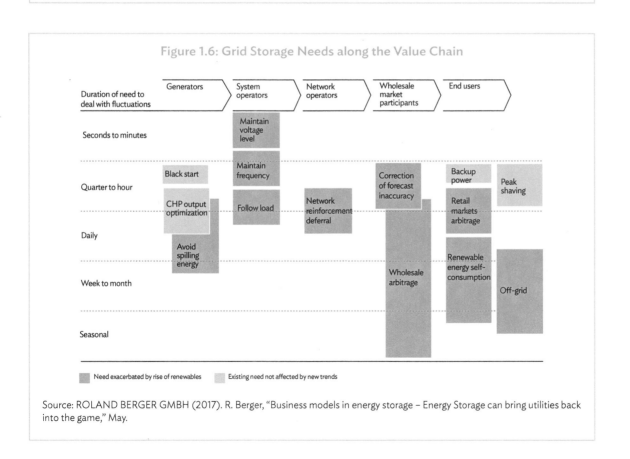

Figure 1.6: Grid Storage Needs along the Value Chain

Source: ROLAND BERGER GMBH (2017). R. Berger, "Business models in energy storage – Energy Storage can bring utilities back into the game," May.

Table 1.1: Discharge Time and Energy-to-Power Ratio of Different Battery Technologies

Discharge Time	Energy-to-Power Ratio	Technologies
Short (seconds to minutes)	Less than 1 (e.g., a capacity of less than 1 kWh for a system with a power of 1 kW)	DLCs, SMES, FES
Medium (minutes to hours)	Between 1 and 10 (e.g., between 1 kWh and 10 kWh for a 1 kW system)	FES, EES such as PbA, Li-ion, and Na–S batteries EES technologies have relatively similar technical features. They have advantages over other technologies in the kW–MW and kWh–MWh range.
Long (days to months)	Considerably greater than 10	Redox flow batteries are situated between storage systems with medium discharge times and those with long discharge times. But their rather low energy density limits the energy-to-power ratio to values between about 5 and 30. DLCs and FES have high power density but low energy density. Li-ion batteries have both high energy density and high power density. This explains the broad range of applications where these batteries are now deployed. Na–S and Na–NiCl2 batteries have higher energy densities than mature battery types such as PbA and Ni–Cd, but they have lower power density than Ni–MH and Li-ion batteries. Metal–air cells have the highest potential in terms of energy density.

DLC = double-layer capacitor, EES = electrochemical energy storage, FES = flywheel energy storage, kW = kilowatt, kWh = kilowatt-hour, MW = megawatt, Li-ion = lithium-ion, Na–NiCl$_2$ = sodium–nickel chloride, Na–S = sodium–sulfur, Ni–Cd = nickel–cadmium, Ni–MH = nickel–metal hydride, PbA = lead–acid, SMES = superconducting magnetic energy storage.

Source: Korea Battery Industry Association, *Energy Storage System Technology and Business Model*, 2017.

1.2 COMPONENTS OF A BATTERY ENERGY STORAGE SYSTEM (BESS)

The various components of a battery energy storage system are shown in the Figure 1.7 Schematic.

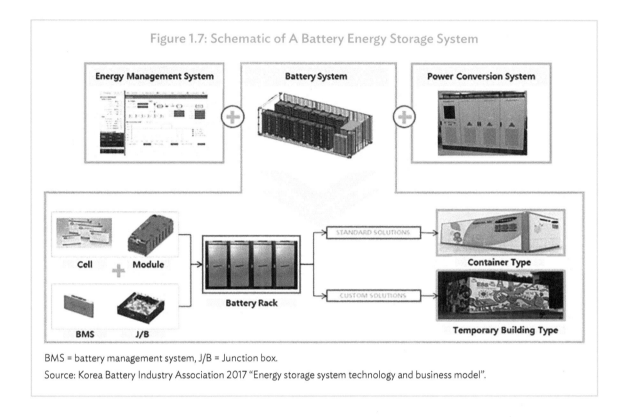

Figure 1.7: Schematic of A Battery Energy Storage System

BMS = battery management system, J/B = Junction box.

Source: Korea Battery Industry Association 2017 "Energy storage system technology and business model".

1.2.1 Energy Storage System Components

ESS components (Figure 1.8) are grouped according to function into battery components, components required for reliable system operation, and grid connection components.

- The battery system consists of the battery pack, which connects multiple cells to appropriate voltage and capacity; the battery management system (BMS); and the battery thermal management system (B-TMS). The BMS protects the cells from harmful operation, in terms of voltage, temperature, and current, to achieve reliable and safe operation, and balances varying cell states-of-charge (SOCs) within a serial connection. The B-TMS controls the temperature of the cells according to their specifications in terms of absolute values and temperature gradients within the pack.

- The components required for the reliable operation of the overall system are system control and monitoring, the energy management system (EMS), and system thermal management. System control and monitoring is general (IT) monitoring, which is partly combined into the overall supervisory control and data acquisition (SCADA) system but may also include fire protection or alarm units. The EMS is responsible for system power flow control, management, and distribution. System thermal management controls all functions related to the heating, ventilation, and air-conditioning of the containment system.

- The power electronics can be grouped into the conversion unit, which converts the power flow between the grid and the battery, and the required control and monitoring components— voltage sensing units and thermal management of power electronics components (fan cooling).

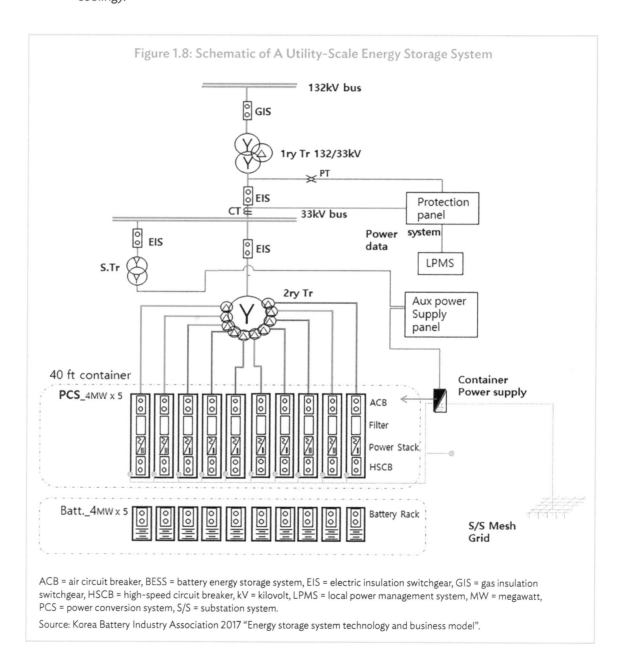

Figure 1.8: Schematic of A Utility-Scale Energy Storage System

ACB = air circuit breaker, BESS = battery energy storage system, EIS = electric insulation switchgear, GIS = gas insulation switchgear, HSCB = high-speed circuit breaker, kV = kilovolt, LPMS = local power management system, MW = megawatt, PCS = power conversion system, S/S = substation system.

Source: Korea Battery Industry Association 2017 "Energy storage system technology and business model".

1.2.2   Grid Connection for Utility-Scale BESS Projects

Figure 1.9 gives an overview of grid connection topologies for utility-scale BESS, which typically consist of multiple battery packs and inverter units, all adding up to the total system energy and power. Power electronics units dedicated to individual battery packs can be installed (Figure 1.9a) or the battery packs can be connected in parallel to a common direct-current (DC) bus (Figure 1.9b). Figure 1.9c shows an example of grid connection to a low-voltage level, and Figure 1.9d, connection to higher grid levels via a transformer.

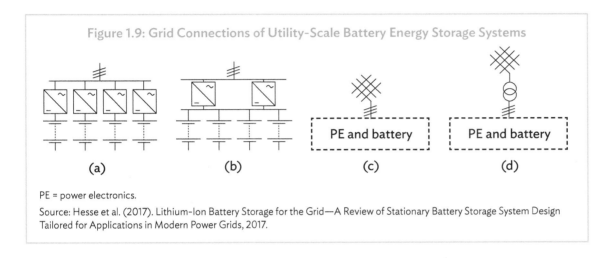

Figure 1.9: Grid Connections of Utility-Scale Battery Energy Storage Systems

(a)   (b)   (c)   (d)

PE = power electronics.

Source: Hesse et al. (2017). Lithium-Ion Battery Storage for the Grid—A Review of Stationary Battery Storage System Design Tailored for Applications in Modern Power Grids, 2017.

1.3 BATTERY CHEMISTRY TYPES

1.3.1   Lead–Acid (PbA) Battery

This type of secondary cell is widely used in vehicles and other applications requiring high values of load current. Its main benefits are low capital costs, maturity of technology, and efficient recycling (Tables 1.2, 1.3, and 1.4).

Table 1.2: Advantages and disadvantages of lead–acid batteries

Advantages	Disadvantages
Low-cost and simple manufacture	Low specific energy; poor weight-to-energy ratio
Low cost per watt-hour	Slow charging: Fully saturated charge takes 14–16 hours
High specific power, capable of high discharge currents	Need for storage in charged condition to prevent sulfation
Good performance at low and high temperatures	Limited cycle life; repeated deep-cycling reduces battery life
No block-wise or cell-wise BMS required	Watering requirement for flooded type
	Transportation restrictions for flooded type
	Adverse environmental impact

BMS = battery management system.

Source: Battery University (2018). "BU-201: How does the Lead Acid Battery Work?," 31 05 2018. [Online]. Available: http://batteryuniversity.com/learn/article/lead_based_batteries

Table 1.3: Types of lead-acid batteries

Type	Description
Sealed, or maintenance-free	First appeared in the mid-1970s. Engineers deemed the term "sealed lead–acid" a misnomer because lead–acid batteries cannot be totally sealed. To control venting during stressful charge and rapid discharge, valves have been added to allow the release of gases if pressure builds up.
Starter	Designed to crank an engine with a momentary high-power load lasting a second or so. For its size, the battery delivers high currents, but it cannot be deep-cycled.
Deep-cycle	Built to provide continuous power for wheelchairs, golf carts, and forklifts, among others. This battery is built for maximum capacity and a reasonably high cycle count.

Source: Battery University (2018a). "BU-201: How does the Lead Acid Battery Work?," 31 05 2018. [Online].
Available: http://batteryuniversity.com/learn/article/lead_based_batteries.

Table 1.4: Uses of lead–acid batteries

Type of Lead–Acid Battery	Uses
Sealed lead–acid (SLA)	Small UPS, emergency lighting, and wheelchairs. Because of its low price, dependable service, and low maintenance requirement, the SLA remains the preferred choice for health care in hospitals and retirement homes.
Valve-regulated lead–acid (VRLA)	Power backup for cellular repeater towers, internet hubs, banks, hospitals, airports, and others
Absorbent glass mat (AGM)	Starter battery for motorcycles, start–stop function for micro-hybrid cars, as well as marine vehicles and RVs that need some cycling

RV = recreational vehicle, UPS = uninterruptible power supply.

Source: Korea Battery Industry Association, *Energy Storage System Technology and Business Model*, 2017.

1.3.2   Nickel–Cadmium (Ni–Cd) Battery

A nickel-cadmium battery (Ni-Cd) is a rechargeable battery used for portable computers, drills, camcorders, and other small battery-operated devices requiring an even power discharge (Table 1.5).

Table 1.5: Advantages and disadvantages of nickel–cadmium Batteries

Advantages	Disadvantages
Rugged, high cycle count with proper maintenance	Relatively low specific energy compared with newer systems
Only battery that can be ultra-fast-charged with little stress	Memory effect; needs periodic full discharge and can be rejuvenated
Good load performance; forgiving if abused	
Long shelf life; can be stored in a discharged state, needing priming before use	Cadmium is a toxic metal; cannot be disposed of in landfills
Simple storage and transportation; not subject to regulatory control	High self-discharge; needs recharging after storage
Good low-temperature performance	Low cell voltage of 1.20 V requires many cells to achieve high voltage
Economical pricing: Ni–Cd has the lowest cost per cycle	
Availability in a wide range of sizes and performance options	

Ni–Cd = nickel–cadmium, V = volt.

Source: Battery University (2018b). "BU-203: Nickel-based Batteries," 31 05 2018. [Online]. Available: http://batteryuniversity.com/

learn/article/nickel_based_batteries.

1.3.3   Nickel–Metal Hydride (Ni–MH) Battery

The Ni–MH battery combines the proven positive electrode chemistry of the sealed Ni–Cd battery with the energy storage features of metal alloys developed for advanced hydrogen energy storage concepts (Moltech Power Systems 2018).

Ni–MH batteries outperform other rechargeable batteries, and have higher capacity and less voltage depression (Table 1.6).

The Ni–MH battery currently finds widespread application in high-end portable electronic products, where battery performance parameters, notably run time, are major considerations in the purchase decision.

Table 1.6: Advantages and Disadvantages of Nickel–Metal Hydride Batteries

Advantages	Disadvantages
Energy density, which can be translated into either long run times or reduction in the space needed for the battery	Limited service life: If repeatedly deep-cycled, especially at high load currents, performance starts to deteriorate after 200–300 cycles. Shallow, rather than deep, discharge cycles are preferred.
Elimination of the constraints imposed on battery manufacture, usage, and disposal because of concerns over cadmium toxicity	Limited discharge current: Although a Ni–MH battery is capable of delivering high discharge currents, repeated discharge with high load currents reduces the battery's cycle life. Best results are achieved with load currents of 0.2–0.5 C (one-fifth to one-half of the rated capacity).
Simplified incorporation into products currently using nickel–cadmium batteries because of the many design similarities between the two chemistries	Need for a more complex charge algorithm: The Ni–MH generates more heat during charge and requires a longer charge time than the Ni–Cd. The trickle charge is critical and must be controlled carefully.
Greater service advantage over other primary battery types at extreme low-temperature operation (–20°C)	High self-discharge: The Ni–MH has about 50% higher self-discharge compared with the Ni–Cd. New chemical additives improve the self-discharge, but at the expense of lower energy density.

C(-rate) = measure of the rate at which a battery is discharged relative to its maximum capacity, Ni–Cd = nickel–cadmium, Ni–MH = nickel–metal hydride.
Source: M. P. systems, "NiMH Technology," 2018. [Online]. Available: https://www.tayloredge.com/.../Batteries/Ni-MH_Generic.pdf

1.3.4   Lithium-Ion (Li-Ion) Battery

Li-ion battery chemistries have the highest energy density and are considered safe. No memory or scheduled cycling is required to prolong battery life. Li-Ion batteries are used in electronic devices such as cameras, calculators, laptop computers, and mobile phones, and are increasingly being used for electric mobility. Their advantages and disadvantages are summarized in Table 1.7 and the various types of such batteries are differentiated in Table 1.8.

Table 1.7: Advantages and disadvantages of lithium-ion batteries

Advantages	Disadvantages
High specific energy and high load capabilities with power cells	Need for protection circuit to prevent thermal runaway if stressed
Long cycle and extended shelf-life; maintenance-free	Degradation at high temperature and when stored at high voltage
High capacity, low internal resistance, good coulombic efficiency	Impossibility of rapid charge at freezing temperatures ($<0°C$, $<32°F$)
Simple charge algorithm and reasonably short charge times	Need for transportation regulations when shipping in larger quantities

Source: Korea Battery Industry Association 2017 "Energy storage system technology and business model".

Table 1.8: Types of lithium-ion batteries

Type	Description
Lithium cobalt oxide ($LiCoO_2$)	The battery consists of a cobalt oxide cathode and a graphite carbon anode. The cathode has a layered structure. During discharge, lithium ions move from the anode to the cathode. The flow reverses on charge. The drawback of Li–cobalt batteries is their relatively short life span, low thermal stability, and limited load capabilities.
Lithium manganese oxide ($LiMn_2O_4$)	Li-ion with manganese spinel was first published in the *Materials Research Bulletin* in 1983. The architecture forms a three-dimensional spinel structure that improves ion flow on the electrode, resulting in lower internal resistance and improved current handling. A further advantage of the spinel structure is high thermal stability and enhanced safety, but the cycle and calendar life are limited.
Lithium nickel manganese cobalt oxide ($LiNiMnCoO_2$, or NMC)	One of the most successful Li-ion systems is a cathode combination of nickel–manganese–cobalt (NMC). Similar to Li–manganese, these systems can be tailored to serve as energy cells or power cells.
Lithium iron phosphate ($LiFePO_4$)	In 1996, the University of Texas (and other contributors) discovered phosphate as cathode material for rechargeable lithium batteries. Li–phosphate offers good electrochemical performance with low resistance. This is made possible with nanoscale phosphate cathode material. The key benefits are high current rating and long cycle life, besides good thermal stability, enhanced safety, and tolerance if abused.
Lithium titanate ($Li_4Ti_5O_{12}$)	Batteries with lithium titanate anodes have been known since the 1980s. Li–titanate replaces the graphite in the anode of a typical lithium-ion battery and the material forms into a spinel structure. The cathode can be lithium manganese oxide or NMC. Li–titanate has a nominal cell voltage of 2.40 V, can be fast-charged, and delivers a high discharge current of 10 C, or 10 times the rated capacity. The cycle count is said to be higher than that of a regular Li-ion. Li–titanate is safe, has excellent low-temperature discharge characteristics, and obtains a capacity of 80% at $-30°C$ ($-22°F$).

C(-rate) = measure of the rate at which a battery is discharged relative to its maximum capacity, Li–cobalt = lithium–cobalt, Li-ion = lithium-ion, Li–phosphate = lithium–phosphate, Li–titanate = lithium–titanate, V = volt.

Source: Battery University (2018c). "BU-205: Types of Lithium-ion," 31 05 2018. [Online]. http://batteryuniversity.com/learn/article/types_of_lithium_ion.

1.3.5 Sodium–Sulfur (Na–S) Battery

The Na–S battery or liquid metal battery is a type of molten metal battery constructed from sodium and sulfur. It exhibits a high energy density, high efficiency of charge and discharge (89%–92%), and a long cycle life, and is fabricated from inexpensive materials (Table 1.9).

However, because of its high operating temperatures of 300°C–350°C and the highly corrosive nature of sodium polysulfides, such cells are primarily used for large-scale nonmobile applications such as electricity grid energy storage (China News Service, n.d.).

Table 1.9: Advantages and disadvantages of sodium–sulfur batteries

Advantages	Disadvantages
Low-cost potential: Inexpensive raw materials and sealed, no-maintenance configuration	Need to be operated above 300°C
High cycle life; liquid electrodes	Highly reactive nature of metallic sodium (part of the material used in construction), which is combustible when exposed to water
Good energy and power density: Low-density active materials, high cell voltage	Extra cost of constructing the enclosing structure to prevent leakage
Flexible operation: Cells functional over a wide range of conditions (rate, depth of discharge, temperature)	Stringent operation and maintenance requirements
High energy efficiency: 100% coulombic-efficient, reasonable resistance	
Insensitivity to ambient conditions: Sealed, high-temperature systems	
State-of-charge identification: Voltage rise and top-of-charge and end-of-discharge	

Source: "Advanced Thin Film Sodium Sulfur Battery," [Online]. Available: en.escn.com.cn/Tools/download.ashx?id=131.

1.3.6 Redox Flow Battery (RFB)

RFBs are charged and discharged by means of the oxidation–reduction reaction of ions of vanadium or the like.

They have excellent characteristics: a long service life with almost no degradation of electrodes and electrolytes, high safety due to their being free of combustible materials, and availability of operation under normal temperatures (Table 1.10). Table 1.11 presents the types of vanadium redox batteries.

Table 1.10: Advantages and disadvantages of redox flow batteries

Advantages	Disadvantages
Long service life: RFBs have a system endurance period of 20 years, with an unlimited number of charge and discharge cycles available without degradation. In addition, the electrolytes can be used semipermanently.	Complexity: RFB systems require pumps, sensors, flow and power management, and secondary containment vessels.
Versatility: With the output and the capacity of a battery capable of being designed independently of each other, RFBs allow flexible design. In addition, the batteries allow a single system to address both short and long periods of output variation, enabling cost-effective power generation.	Low energy density: The energy densities of RFBs are usually low compared with those of other types of batteries.
High safety: RFBs are capable of operating under normal temperatures and are composed of noncombustible or flame-retardant materials. The possibility of a fire with the batteries is extremely low.	

RFB = redox flow battery.

Source: Sumitomo Electric Industries Ltd. (n.d.).
Sumitomo electric, "Redox Flow Battery," [Online]. Available: http://global-sei.com/products/redox/.

Table 1.11: Types of vanadium redox batteries

Type	Description
Vanadium redox battery (VRB)	VRBs use two vanadium electrolytes ($V2+/V3+$ and $V4+/V5+$), which exchange hydrogen ions (H+) through a membrane.
Polysulfide–bromine battery (PSB)	Sodium sulfide (Na_2S_2) and sodium tribromide ($NaBr_3$) are used as electrolytes. The sodium ions (Na+) pass through the membrane during the charging or discharging process.
Zinc–bromine (Zn–Br) battery	Solutions of zinc and a complex bromine compound are used as electrodes.

Source: Sumitomo Electric Industries Ltd. (n.d.).
Sumitomo electric, "Redox Flow Battery," [Online]. Available: http://global-sei.com/products/redox/.

2 BUSINESS MODELS FOR ENERGY STORAGE SERVICES

2.1 OWNERSHIP MODELS

There are various business models through which energy storage for the grid can be acquired as shown in Table 12. According to Abbas, A. et. al., these business models include service-contracting without owning the storage system to "outright purchase of the BESS. The needs and preference of the service user will determine the specific option to be chosen. This chapter presents the general principles for owning and operating BESS through various options.[1]

Table 2.1: Energy storage ownership models

Wholesale	Substation	End-Use Customer
Utility-owned	Utility-owned	Customer–owned
IPP-owned	-Grid asset	ESCO (with aggregator)–owned
Supplier-/Vendor-owned	-Smart-grid asset	IPP–owned
	IPP-owned	Utility (LSE)–owned
	ESCO-owned	Part of utility program
	IPP/LSE contract for grid support services	

ESCO = energy service company, IPP = independent power producer, LSE = load-serving entity.

Source: Korea Battery Industry Association 2017 "Energy storage system technology and business model".

2.1.1 Third-Party Ownership

In this option, the storage system is owned, operated, and maintained by a third-party, which provides specific storage services according to a contractual arrangement. This process is very similar to power purchase agreements signed with independent power producers. Third-Party ownership contracts similar to those offered to thermal power plants, typically lasting for 20–25 years, generally include the following key terms:

- The off-taker holds the dispatch rights for charging and discharging the energy storage system (ESS).
- The seller earns a fixed capacity payment ($/kW-month) and a variable payment for operation and maintenance (O&M) per MWh delivered ($/MWh).
- In return for the capacity payment, the seller provides assurance of a specified degree of availability of the plant.
- The seller provides an efficiency guarantee.

---

[1] A. A. Akhil, G. Huff, A. Currier, B. Kaun, D. Rastler, SB Chen, A. Cotter, D. Bradshaw, and W. Gauntlett, "Electricity Storage Handbook," https://prod.sandia.gov/techlib-noauth/access-control.cgi/2015/151002.pdf, (January 2015).

2.1.2   Outright Purchase and Full Ownership

In outright purchase and full ownership, the wide difference in size and functionality between pumped hydro and compressed-air energy storage (CAES) technologies, on the one hand, and batteries and flywheels, on the other, creates a clear distinction between their procurement and installation processes.[2]

2.1.3   Electric Cooperative Approach to Energy Storage Procurement

While investor-owned utilities (IOUs) and electric cooperatives often have similar electricity storage needs, they differ in ownership, governance, and financial structure, as well as in infrastructure and customer demographics. These differences could affect their approach to capital asset ownership.

IOUs operate for profit, are funded by their investors and by public sector and bank borrowings, and are governed by and generate profits for their shareholders, who may not live in the IOU service area.

Co-op utilities, on the other hand, are not-for-profit entities existing to serve their owner-members, all of whom live in the co-op service area. Co-ops grew out of the decision of residents of localities without electricity access—generally those with few and widely dispersed consumers, not enough to produce a profit for investor-owned power companies—to get that access themselves through their own power companies. These utilities use loans, grants, and private financing for operation, maintenance, and modernization. Surplus revenue goes back to the members, depending on their electricity consumption (patronage dividends). Members have voting rights and a hand in setting policies and running the business.

Co-op utilities are of two types. Distribution co-ops deliver electricity to their owner-members, while generation and transmission (G&T) co-ops own and operate generation assets and sell bulk power to distribution co-ops under all-requirements contracts, thus essentially agreeing to be the single-source provider of their power needs.

The entity benefiting from the ESS is therefore an important consideration in the choice between service acquisition options.

---

[2]   A. A. Akhil, et. al. 2015.

2.2   FACTORS AFFECTING THE VIABILITY OF BESS PROJECTS

The economic and financial viability of BESS projects depends on several factors (Table 2.2).

Table 2.2: Key factors affecting the viability of battery energy storage system projects

Factor	Impact on Project Viability
Cost of storage	Battery costs, while falling, are still the most significant driver of project viability. Costs depend on the MW/MWh ratio of the battery. The terminal value at the end of the project's economic life also has a bearing, with a higher terminal value improving project economics.
Network reinforcement cost	Higher conventional network reinforcement costs increase the value of deploying storage as an alternative, improving project economics (and vice versa) for DNOs directly and for third-party projects with a contract for peak shaving with a DNO.
Commercial services	Increased access to and higher value from the provision of commercial services (for example, ancillary service markets, the wholesale market, the capacity market) increase project revenue streams, improving project economics (and vice versa). It is generally accepted that value streams will need to be stacked to increase the economic viability of BESS projects (see Figure 10).
Policy developments	Removing barriers to storage or creating a more favorable environment for investment enhances the realizable value of a project, improving project economics (and vice versa).

BESS = battery energy storage system, DNO = distribution network operator, MW = megawatt, MWh = megawatt-hour.

Source: Korea Battery Industry Association 2017 "Energy storage system technology and business model".

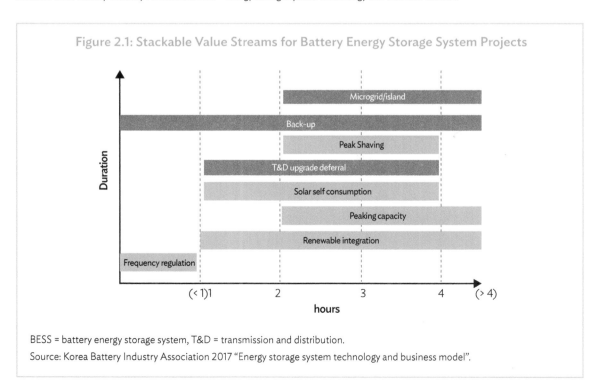

Figure 2.1: Stackable Value Streams for Battery Energy Storage System Projects

BESS = battery energy storage system, T&D = transmission and distribution.

Source: Korea Battery Industry Association 2017 "Energy storage system technology and business model".

2.3 FINANCIAL AND ECONOMIC ANALYSIS

Based on the Guidelines for the Economic Analysis of Projects by the ADB, project economic analysis and financial evaluation both involve identifying project benefits and costs during the years in which they occur and converting all future cash flows into their present value by means of discounting. Both analyses generate net present value (NPV) and internal rate of return (IRR) indicators.[3]

However, the perspectives and objectives of the two analyses differ:

- Financial evaluation assesses the ability of the project to generate adequate incremental cash flows for the recovery of financial costs (capital and recurrent costs) without external support.
- Project economic analysis (Figure 2.2) assesses whether a project is economically viable for the country.

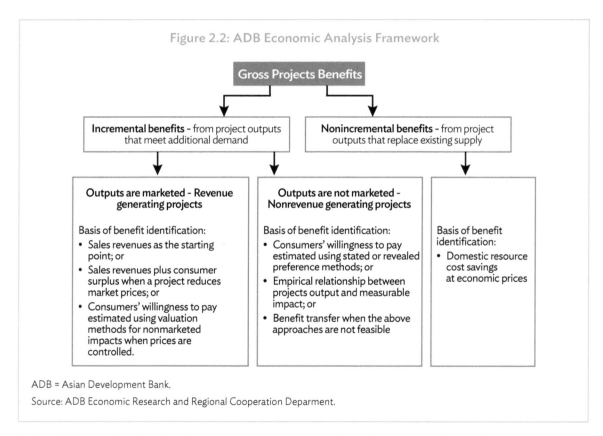

Figure 2.2: ADB Economic Analysis Framework

Gross Projects Benefits

Incremental benefits - from project outputs that meet additional demand

Nonincremental benefits - from project outputs that replace existing supply

Outputs are marketed - Revenue generating projects

Basis of benefit identification:
- Sales revenues as the starting point; or
- Sales revenues plus consumer surplus when a project reduces market prices; or
- Consumers' willingness to pay estimated using valuation methods for nonmarketed impacts when prices are controlled.

Outputs are not marketed - Nonrevenue generating projects

Basis of benefit identification:
- Consumers' willingness to pay estimated using stated or revealed preference methods; or
- Empirical relationship between projects output and measurable impact; or
- Benefit transfer when the above approaches are not feasible

Basis of benefit identification:
- Domestic resource cost savings at economic prices

ADB = Asian Development Bank.
Source: ADB Economic Research and Regional Cooperation Deparment.

In identifying project benefits for economic analysis, two distinctions are particularly important:

- The first is whether the benefits are derived from incremental or from non-incremental output.
- The second distinction is whether project output is sold in markets, and whether there are market prices that can be used as the starting point for valuing project benefits.

---

3   ADB. 2015. *Guidelines for the Economic Analysis of Projects.* Manila.

2.3.1 Criteria for the Economic Analysis of BESS Projects

Discount rate. The expected net present value (ENPV) and the economic internal rate of return (EIRR) should be calculated for all projects in which benefits can be valued. The general criterion for accepting a project is achieving a positive ENPV discounted at the minimum required EIRR or achieving the minimum required EIRR of 9%.[4]

Shadow exchange rate. Multiplying output and input values measured at world prices and converted at the official exchange rate using the shadow exchange rate factor (SERF), while leaving those at domestic prices unadjusted, brings the former to a common base of measurement with the latter, which is in the currency of the borrowing country at its domestic price level.[5]

2.3.2 Key Assumptions in the Cost–Benefit Analysis of BESS Projects

Battery cell prices. The feasibility analysis can be extremely sensitive to the price assumptions for the terminal value of used cells and the future cost of replacement cells. Lithium-ion cell prices are expected to continue falling over the next few years as manufacturing capacity ramps up (Figures 2.3 and 2.4).

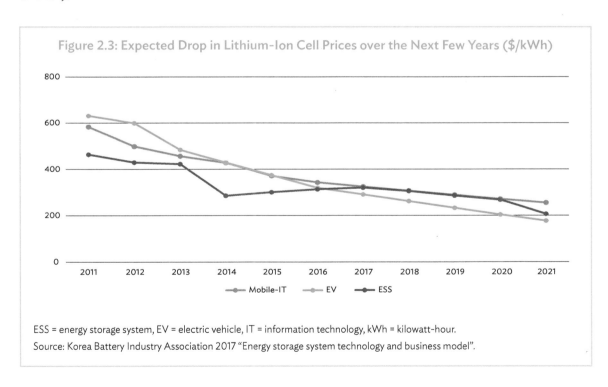

Figure 2.3: Expected Drop in Lithium-Ion Cell Prices over the Next Few Years ($/kWh)

ESS = energy storage system, EV = electric vehicle, IT = information technology, kWh = kilowatt-hour.

Source: Korea Battery Industry Association 2017 "Energy storage system technology and business model".

---

4   ADB, 2017.
5   The ratio of the shadow exchange rate (SER) to the official exchange rate.

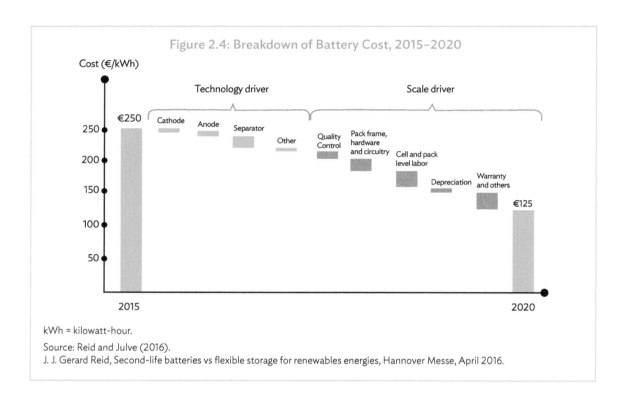

Figure 2.4: Breakdown of Battery Cost, 2015–2020

kWh = kilowatt-hour.
Source: Reid and Julve (2016).
J. J. Gerard Reid, Second-life batteries vs flexible storage for renewables energies, Hannover Messe, April 2016.

Benchmark capital costs for a fully installed grid-scale energy storage system. A continuous fall in the capital cost of building grid-scale ESSs is also projected (Figure 2.5).

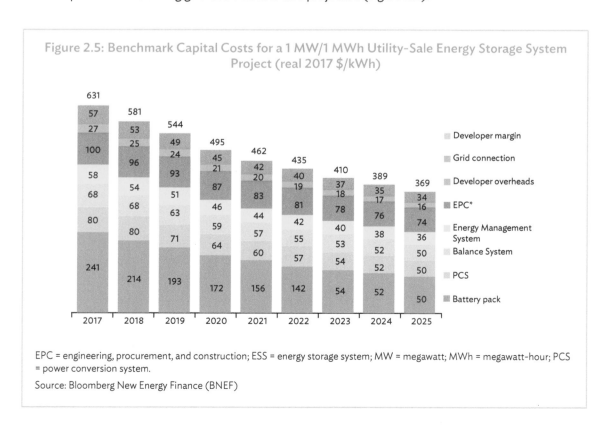

Figure 2.5: Benchmark Capital Costs for a 1 MW/1 MWh Utility-Sale Energy Storage System Project (real 2017 $/kWh)

EPC = engineering, procurement, and construction; ESS = energy storage system; MW = megawatt; MWh = megawatt-hour; PCS = power conversion system.
Source: Bloomberg New Energy Finance (BNEF)

Benchmark capital costs for a fully installed residential energy storage system. The capital cost of residential ESS projects are similarly foreseen to drop over the next few years (Figure 2.6).

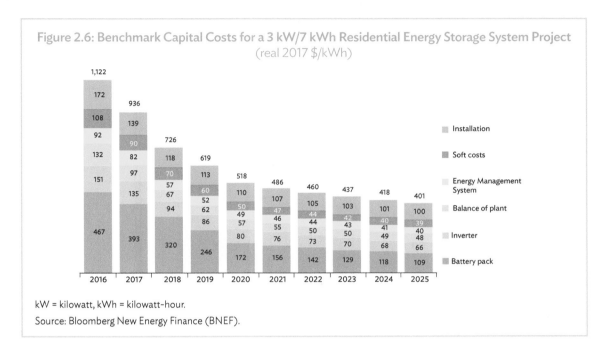

Figure 2.6: Benchmark Capital Costs for a 3 kW/7 kWh Residential Energy Storage System Project (real 2017 $/kWh)

kW = kilowatt, kWh = kilowatt-hour.

Source: Bloomberg New Energy Finance (BNEF).

Battery life. The useful life (in cycles) of a battery depends on two factors: cell chemistry and aging. Cell chemistry includes, anode and cathode materials, cell capacity (in ampere-hours, or amp-hours), energy density (in watt-hours per liter), and energy-to-power ratio. Table 2.3 compares the cell chemistry of different types of lithium-ion batteries.

Table 2.3: Comparison of Different Lithium-Ion Battery Chemistries

Cathode	Anode	Energy Density (watt-hours/kg)	Number of Cycles
LFP	Graphite	85–105	200–2,000
LMO	Graphite	140–180	800–2,000
LMO	LTO	80–95	2,000–25,000
LCO	Graphite	140–200	300–800
NCA	Graphite	120–160	800–5,000
NMC	Graphite, silicone	120–140	800–2,000

kg = kilogram, LCO = lithium–cobalt oxide, LFP = lithium–iron–phosphate, LMO = lithium–manganese oxide, LTO = lithium–titanate oxide, NCA = nickel–cobalt–aluminum oxide, NMC = nickel–manganese–cobalt.

Source: IRENA (2015).

Aging is due to the fading of active materials caused by the charge and discharge cycles. Batteries discharged below a 20% SOC—more than 80% depth-of-discharge (DOD)—age faster. For example, a 7 watt-hour lithium–nickel–manganese–cobalt (lithium–NMC) battery cell can perform over 50,000 cycles at 10% cycle depth, yielding a lifetime energy throughput (the total amount of energy charged and discharged from the cell) of 35 kWh. But the same cell cycled at 100% cycle depth can perform only 500 cycles, yielding a lifetime energy throughput of only 3.5 kWh. (Figure 2.7 shows the lifetime energy throughput of lithium–iron–phosphate batteries at different cycle depths.)

When carrying out financial and economic analysis, the maximum DOD should be limited to 80% to prolong battery life, and replacement cell purchase should be considered when the battery reaches 80% of its useful life (in cycles) to avoid degradation of performance.

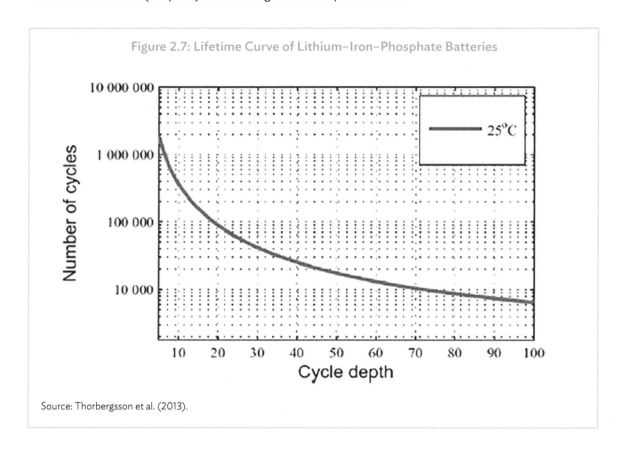

Figure 2.7: Lifetime Curve of Lithium–Iron–Phosphate Batteries

Source: Thorbergsson et al. (2013).

3 GRID APPLICATIONS OF BATTERY ENERGY STORAGE SYSTEMS

3.1 SCOPING OF BESS USE CASES

The services provided by batteries can be divided into groups representing the primary stakeholders (Table 3.1).

Table 3.1: Energy Storage Use Case Applications, by Stakeholder

Stakeholder	BESS Services (Use Cases)
Network owners	Peak-load management or investment deferral in system reinforcement
Network operators	Ancillary services such as frequency regulation or voltage support
"Behind-the-meter" customer services	Increased self-consumption of solar PV, backup power, peak-time charge reduction, etc.

BESS = battery energy storage system, PV = photovoltaic.
Source: Korea Battery Industry Association 2017 "Energy storage system technology and business model."

A major advantage provided by battery energy storage is flexibility in addressing the full range of active and reactive power needs (Figure 3.2). The Rocky Mountain Institute translated this capability into discrete grid services at the generation, transmission, and distribution levels of the electricity system.[6]

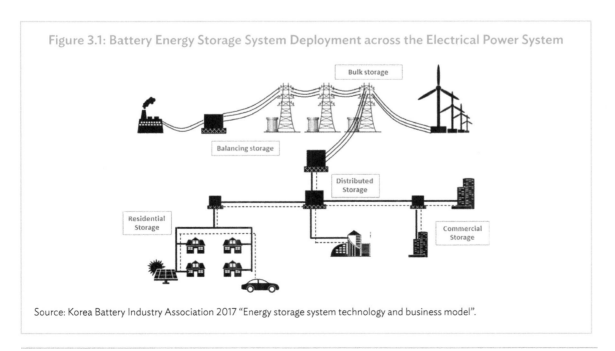

Figure 3.1: Battery Energy Storage System Deployment across the Electrical Power System

Source: Korea Battery Industry Association 2017 "Energy storage system technology and business model".

---

6   G. Fitzgerald, J. Mandel, J. Morris, and H. Touti. 2015. *The Economics of Battery Energy Storage*. Rocky Mountain Institute. https://rmi. org/insight/economics-battery-energy-storage/.

3.2   GENERAL GRID APPLICATIONS OF BESS

BESS grid applications are summarized in Table 3.2 together with the technical factors involved according to the Electricity Storage Handbook.[7]

Table 3.2: Technical Considerations for Grid Applications of Battery Energy Storage Systems

Grid Application	Technical Considerations
Electric energy time-shift (arbitrage) Electric energy time-shift involves purchasing inexpensive electric energy, available during periods when prices or system marginal costs are low, to charge the storage system so that the stored energy can be used or sold at a later time when the price or costs are high. Alternatively, storage can provide similar time-shift duty by storing excess energy production, which would otherwise be curtailed, from renewable sources such as wind or photovoltaic.	Storage system size range: 1 – 500 MW Target discharge duration range: < 1 hour Minimum cycles/year: 250 +
Electric supply capacity Depending on the circumstances in a given electric supply system, energy storage could be used to defer or reduce the need to buy new central station generation capacity or purchasing capacity in the wholesale electricity marketplace.	Storage system size range: 1–500 MW Target discharge duration range: 2–6 hours Minimum cycles/year: 5–100
Regulation Regulation is one of the ancillary services for which storage is especially well suited. It involves managing interchange flows with other control areas to match closely the scheduled interchange flows and momentary variations in demand within the control area. The primary reason for including regulation in the power system is to maintain the grid frequency.	Storage system size range: 10–40 MW Target discharge duration range: 15 minutes to 1 hour Minimum cycles/year: 250–10,000
Spinning, non-spinning, and supplemental reserves The operation of an electric grid requires reserve capacity that can be called on when some portion of the normal electric supply resources unexpectedly become unavailable. Generally, reserves are at least as large as the single largest resource (e.g., the single largest generation unit) serving the system, and reserve capacity is equivalent to 15%–20% of the normal electric supply capacity.	Storage system size range: 10–100 MW Target discharge duration range: 15 minutes to 1 hour Minimum cycles/year: 20–50
Voltage support Normally, designated power plants are used to generate reactive power (expressed in VAr) to offset reactance in the grid. These power plants could be displaced by strategically placed energy storage within the grid at central locations or by multiple VAr-support storage systems placed near large loads, following the distributed approach. The PCS of the storage systems used for voltage support must be capable of operating at a non-unity power factor, to source and sink reactive power.	Storage system size range: 1–10 MVAr Target discharge duration range: Not applicable Minimum cycles/year: Not applicable
Black start Storage systems provide an active reserve of power and energy within the grid and can be used to energize transmission and distribution lines and provide station power to bring power plants on line after a catastrophic failure of the grid. Storage can provide similar start-up power to larger power plants, if the storage system is suitably sited and there is a clear transmission path to the power plant from the storage system's location.	Storage system size range: 5–50 MW Target discharge duration range: 15 minutes to 1 hour Minimum cycles/year: 10–20

continued on next page

[7]   A. Akhil, et. al., 2015.

Table 3.2 *continued*

Grid Application	Technical Considerations
Load following/Ramping up of renewables Load following is characterized by power output that generally changes as often as every several minutes. The output changes in response to the changing balance between electric supply and load within a specific region or area. Output variation is a response to changes in system frequency, timeline loading, or the relation of these to each other, and occurs as needed to maintain the scheduled system frequency or established interchange with other areas within predetermined limits.	Storage system size range: 1–100 MW Target discharge duration range: 15 minutes to 1 hour Minimum cycles/year: Not applicable
Transmission upgrade deferral Transmission upgrade deferral involves delaying utility investments in transmission system upgrades, by using relatively small amounts of storage, or in some cases avoiding such investments entirely. Consider a transmission system with peak electric loading that is approaching the system's load-carrying capacity (design rating). In some cases, installing a small amount of energy storage downstream from the nearly overloaded transmission node could defer the need for the upgrade for a few years.	Storage system size range: 10–100 MW Target discharge duration range: 2–8 hours Minimum cycles/year: 10–50
Transmission congestion relief Transmission congestion occurs when energy from dispatched power plants cannot be delivered to all or some loads because of inadequate transmission facilities. When transmission capacity additions do not keep pace with the growth in peak electric demand, the transmission systems become congested. Electricity storage can be used to avoid congestion-related costs and charges, especially if the costs become onerous because of significant transmission system congestion.	Storage system size range: 1–100 MW Target discharge duration range: 1–4 hours Minimum cycles/year: 50–100
Distribution upgrade deferral and voltage support Distribution upgrade deferral involves using storage to delay or avoid investments that would otherwise be necessary to maintain adequate distribution capacity to serve all load requirements. The deferred upgrade could be a replacement of an aging or overstressed distribution transformer at a substation or the re-conductoring of distribution lines with heavier wire.	Storage system size range: 500 kW–10 MW Target discharge duration range: 1–4 hours Minimum cycles/year: 50–100
Power quality The electric power quality service involves using storage to protect customer on-site loads downstream (from storage) against short-duration events that affect the quality of power delivered to the customer's loads. Some manifestations of poor power quality are the following: • variations in voltage magnitude (e.g., short-term spikes or dips, longer-term surges or sags); • variations in the primary 60-hertz (Hz) frequency at which power is delivered; • low power factor (voltage and current excessively out of phase with each other); • harmonics (the presence of currents or voltages at frequencies other than the primary frequency); and • interruptions in service, of any duration, ranging from a fraction of a second to several seconds.	Storage system size range: 100 kW–10 MW Target discharge duration range: 10 seconds to 15 minutes Minimum cycles/year: 10–200

continued on next page

Table 3.2 *continued*

Grid Application	Technical Considerations
Demand charge management Electricity storage can be used by end users (utility customers) to reduce their overall costs for electric service by reducing their demand during peak periods specified by the utility. To avoid a demand charge, load must be reduced during all hours of the demand charge period, usually a specified period of time (e.g., 11 a.m. to 5 p.m.) and on specified days (most often weekdays).	Storage system size range: 50 kW–10 MW Target discharge duration range: 1–4 hours Minimum cycles/year: 50–500

kW = kilowatt, MW = megawatt, MVAr = megavolt-ampere reactive, PCS = power conversion system, VAr = volt-ampere reactive.

Source: Sandia National Laboratories (2013).

Sandia National Laboratories, "DOE/EPRI 2013 Electricity Storage Handbook in Collaboration with NRECA," DOE, 2013.

3.3 TECHNICAL REQUIREMENTS

3.3.1   Round-Trip Efficiency

Round-trip efficiency takes into consideration energy losses from power conversions and parasitic loads (e.g., electronics, heating and cooling, and pumping) associated with operating the energy storage system. This metric is a key determinant of the cost-effectiveness of energy storage technologies. Among energy storage options, compressed-air energy storage (CAES) has the lowest reported efficiency (40%–55%), and Li-ion batteries have the highest (87%–94%).[8]

For energy storage coupled with photovoltaics, efficiencies of less than 75% are unlikely to be cost-effective.

3.3.2   Response Time

The need for fast response times is expected to be more important for variability-damping than for load-shifting applications, and hence more relevant to utility-scale photovoltaic generation in this evaluation. Passing clouds are the primary source of rapid changes in photovoltaic power output. Solar insolation at a single point can change by more than 60% in seconds.[9]

Changes of this magnitude in power output from utility-scale photovoltaic systems are expected to occur within minutes.

Photovoltaic power output ramp rates were measured over a year at a photovoltaic system in Hawaii that operated at 50% capacity. In that study, only 0.07% of the one-minute ramps were operating at greater than 60% capacity and only 5%, at greater than 10% of that capacity.[10] System operator experience suggests that a response time of seconds would be adequate to dampen short-term variability events of significant magnitude.

---

8   EPRI (Electric Power Research Institute), "Electricity Energy Storage Technology Options: A White Paper Primer on Applications, Costs, and Benefits," Palo Alto, Calif., 2010a.

9   R. Berger, "Business models in energy storage – Energy Storage can bring utilities back into the game," May 2017.

10   J. Johnson. , S. B., E. A., Q. J. and L. C., "Initial Operating Experience of the La Ola 1.2 MW Photovoltaic System, Report SAND2011-8848,," Sandia National Laboratories, 2011

3.3.3 Lifetime and Cycling

As is the case for efficiency, the cost-effectiveness of energy storage is directly related to its operational lifetime. The lifetime of an ESS depends on many factors, including charge and discharge cycling, depth of discharge, and environmental conditions. For any application, maximizing the depth of discharge minimizes the required energy storage capacity. The cycling schedule thus offers the greatest degree of freedom in design. For residential and commercial applications, one or two cycles per day—or 7,300–22,000 lifetime cycles—would be adequate to allow for photovoltaic power shifting and nighttime storage of cheap grid electricity.

Increasing the magnitude or percentage of power spikes that is tolerated without intervention would also reduce the lifetime cycling requirement. An energy storage system that is designed to respond to power spikes that are more than 10% of the photovoltaic nameplate power in a single charge and discharge cycle might experience many more than 100,000 cycles over its lifetime, especially in cloudier locations.

3.3.4 Sizing

Frequency regulation and black start BESS grid applications are sized according to power converter capacity (in MW). These other grid applications are sized according to power storage capacity (in MWh): renewable integration, peak shaving and load leveling, and microgrids.

Sized by power converter capacity [MW]	Sized by power storage capacity [MWh]
Frequency regulation, black start	Renewable integration, peak shaving and/or load leveling, microgrids

Examples

$$BESS\ Capacity\ [MW] = Frequency\ gain\ \left[\frac{MW}{Hz}\right] * Governor\ droop\ [\%] * System\ frequency\ [Hz]$$

e.g., $41.667\frac{MW}{Hz} * 4\% * 60\ Hz = 100\ MW$

$$BESS\ Capacity\ [MWh] = \frac{power\ required\ [MW] * duration\ required\ [h]}{depth\ of\ discharge\ [\%] * battery\ efficiency\ [\%]}$$

e.g., $\frac{100\ MW * 4\ h}{80\% * 95\%} = 263\ MWh$

BESS = battery energy storage system, h = hour, Hz = hertz, MW = megawatt, MWh = megawatt-hour.

3.4   OPERATION AND MAINTENANCE

Timely operation and maintenance of the facility is required to minimize loss of energy yield, damage to property, safety concerns, and disruption of electric power supply (Table 3.3). Maintenance is both preventive and corrective to maximize BESS output and ensure uninterrupted operation.

Table 3.3: Operation and maintenance of battery energy storage systems

Function	Definition
Operation	Monitoring system management, operation status check and repair, data management and reporting
Corrective maintenance	Component and equipment-wise checks and repair, repair work (following expiration of EPC warranty period), verification of repairs, documentation
Environmental management	Vegetation abatement, waste and garbage dumping, battery disposal
Safety management	Protection of the ESS facility against criminal acts such as vandalism, theft, and trespassing
Transmission-line management	Transmission-line check and repair work
Spare parts	Ample storage of on-site spares with suitable safeguards is crucial for meeting the performance guarantees maintaining a higher level of BESS uptime than the plant availability agreement

BESS = battery energy storage system; EPC = engineering, procurement, and construction; ESS = energy storage system.

Source: Korea Battery Industry Association 2017 "Energy storage system technology and business model".

3.5   USE CASES

3.5.1 Frequency Regulation

Frequency regulation is the constant second-by-second adjustment of power to maintain system frequency at the nominal value (50 or 60 Hz) to ensure grid stability (Figure 3.2).

If demand exceeds supply, the system frequency falls, and brownouts and blackouts are likely. If utilities generate more power than consumers demand, the system frequency increases, possibly damaging all connected electrical devices.

Battery energy storage can provide regulating power with sub-second response times (Figure 3.3). This makes it an extremely useful asset for grid-balancing purposes.

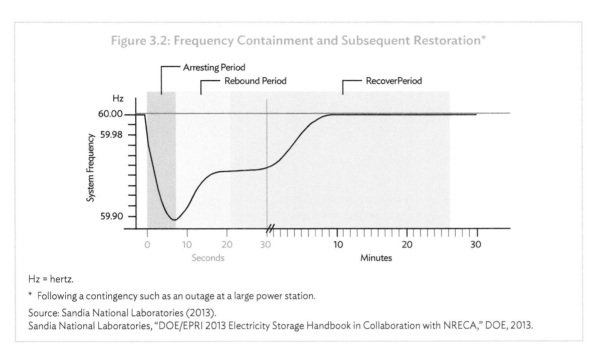

Figure 3.2: Frequency Containment and Subsequent Restoration*

Hz = hertz.

* Following a contingency such as an outage at a large power station.

Source: Sandia National Laboratories (2013).
Sandia National Laboratories, "DOE/EPRI 2013 Electricity Storage Handbook in Collaboration with NRECA," DOE, 2013.

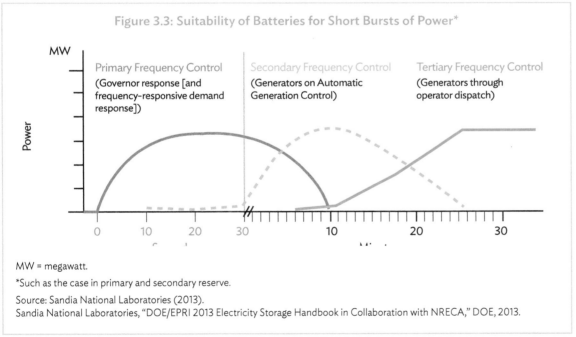

Figure 3.3: Suitability of Batteries for Short Bursts of Power*

MW = megawatt.

*Such as the case in primary and secondary reserve.

Source: Sandia National Laboratories (2013).
Sandia National Laboratories, "DOE/EPRI 2013 Electricity Storage Handbook in Collaboration with NRECA," DOE, 2013.

Real-time data, such as battery SOC, rated power or effective power, and system frequency, must be communicated to the transmission system operator (TSO) using the BESS as part of its pool of assets for frequency regulation.

3.5.2 Renewable-Energy Integration

Future power systems are expected to rely significantly on renewable energy sources (RESs) such as wind and solar. However, the variability and intermittence of solar photovoltaic and wind-power generation present challenges for safe and reliable grid integration:

- Network owners may be penalized for insufficient reinforcement of network capacity to accommodate all network connection requests. In one early case in 2013, the Hawaiian Electric Company was forced to temporarily stop issuing interconnection permits for distributed solar installations.
- Distribution system operators may have trouble with system stability and may be forced to curtail renewable in-feed to avoid over-voltage conditions. TSOs may be forced to hold larger amounts of spinning reserve to compensate for large forecast errors.
- Network reinforcement and curtailment costs are ultimately recovered from end consumers in the form of higher tariffs.

Energy storage provides the power system with flexibility and is very useful in increasing the volume of renewable power that can be safely and securely connected to the grid:

- More grid connections can be made under existing network capacity as surplus power can be stored.
- Smoothing of renewable in-feed reduces forecast errors, and thus the need to procure spinning reserve. Surplus power can be stored at consumers' homes instead of being fed into the grid.
- Higher network capacity utilization reduces the burden on consumers as curtailments are reduced and network reinforcement is minimized.

Solar photovoltaic generation. The overall generation forecast is good, with well-defined peaks. However, the in-feed can be volatile when there is cloud cover, and large variations can occur minute by minute (Figures 3.4 and 3.5).

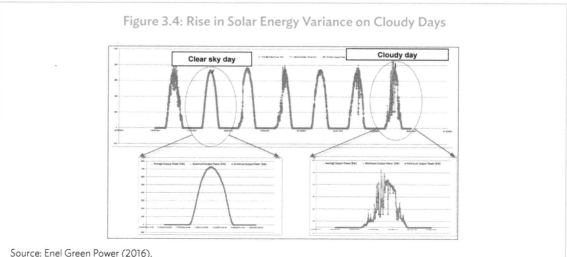

Figure 3.4: Rise in Solar Energy Variance on Cloudy Days

Source: Enel Green Power (2016).
Enel Green Power, "EGP Integrating renewable power plants with energy storage," 7 6 2016. [Online]. Available: http://www.iefe.unibocconi.it/wps/wcm/connect/29b685e1-8c34-4942-8da3-6ab5e701792b/Slides+Lanuzza+7+giugno+2016.pdf?MOD=AJPERES&CVID=lle7w78

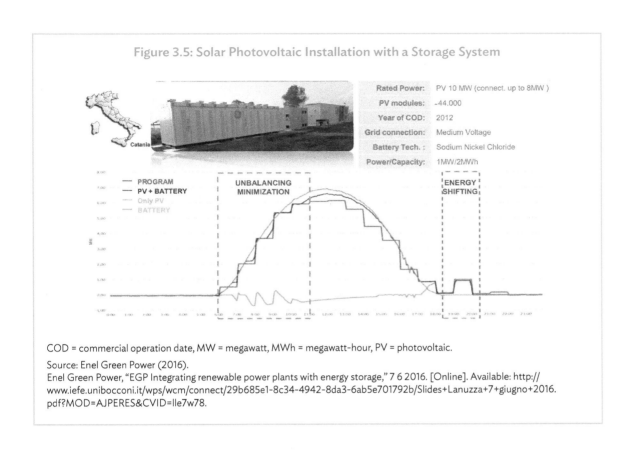

Figure 3.5: Solar Photovoltaic Installation with a Storage System

COD = commercial operation date, MW = megawatt, MWh = megawatt-hour, PV = photovoltaic.

Source: Enel Green Power (2016).
Enel Green Power, "EGP Integrating renewable power plants with energy storage," 7 6 2016. [Online]. Available: http://www.iefe.unibocconi.it/wps/wcm/connect/29b685e1-8c34-4942-8da3-6ab5e701792b/Slides+Lanuzza+7+giugno+2016.pdf?MOD=AJPERES&CVID=lle7w78.

Wind-power generation. The forecast is more challenging than that for solar photovoltaic. Large variations can occur within minutes (Figure 3.6). Wind turbines are usually disabled when wind speed exceeds 25 meters per second, potentially resulting in a large drop in power generation.

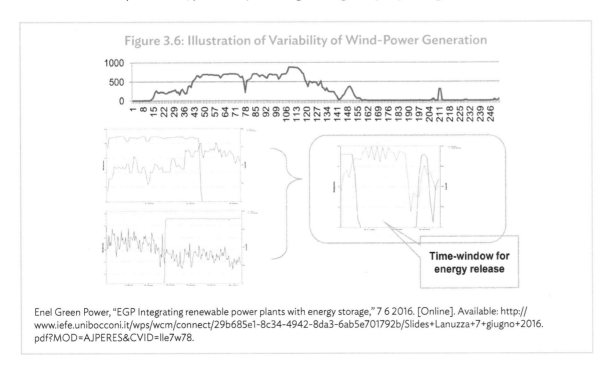

Figure 3.6: Illustration of Variability of Wind-Power Generation

Enel Green Power, "EGP Integrating renewable power plants with energy storage," 7 6 2016. [Online]. Available: http://www.iefe.unibocconi.it/wps/wcm/connect/29b685e1-8c34-4942-8da3-6ab5e701792b/Slides+Lanuzza+7+giugno+2016.pdf?MOD=AJPERES&CVID=lle7w78.

The main challenges presented by wind-power integration are power intermittence, ramp rate, and limited wind-farm output. The BESS can improve wind-energy dispatch by reducing forecast errors and improving the utilization of transmission capacity. The BESS can also be used by the system operator for providing ancillary services to mitigate the variability and uncertainty of wind power on the grid side.

3.5.3   Peak Shaving and Load Leveling

Peak shaving. The reduction of electric power demand during times when network capacity is stressed is known as peak shaving (Figure 3.7). Peak shaving helps defer investments in network expansion or network reinforcement.

Peak shaving also helps the utility meet demand without having to ramp up expensive peaking generators. In the long run, peak shaving allows investment deferral in the building of new power plants. Customers that install on-site power generation share in those savings by receiving reduced power tariffs or in some cases capacity payments (such as in the United Kingdom [UK]).

Peak demand generally occurs in the afternoons:

- In the Republic of Korea: between 3 p.m. and 8 p.m. in July and August, the time of maximum power demand for air-conditioning and other nonindustrial usage.
- In the UK: between 5 p.m. and 6:30 p.m. from November to February, the time of maximum heating demand.

Peak-shaving generators use special equipment to monitor the electric grid and start up quickly. This equipment provides the added benefit of backup power in case of rolling backups and grid outages (Figure 3.7).

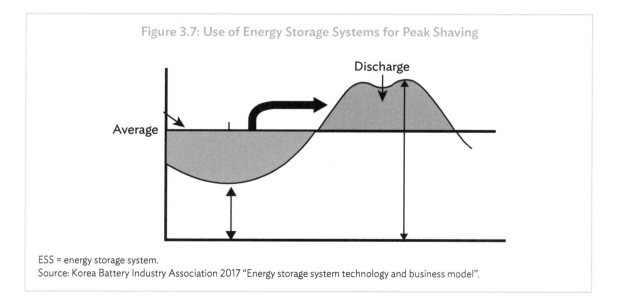

Figure 3.7: Use of Energy Storage Systems for Peak Shaving

ESS = energy storage system.
Source: Korea Battery Industry Association 2017 "Energy storage system technology and business model".

Load leveling. Load leveling, on the other hand, refers to the process of shifting demand away from peak hours to off-peak hours (Figure 3.8). Load leveling can be performed by introducing time-of-use tariffs, whereby consumers are incentivized to shift consumption to hours of the day when tariffs are low. Behind-the-meter energy storage allows for load leveling (from the utility perspective) without any changes to the consumer load profile.

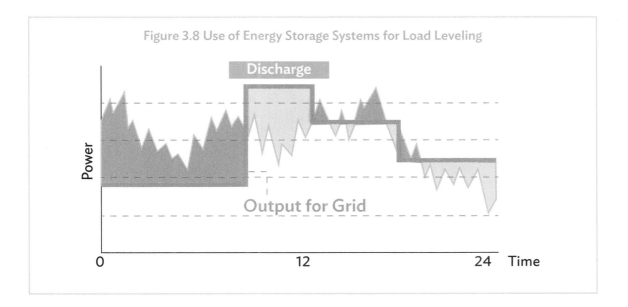

Figure 3.8 Use of Energy Storage Systems for Load Leveling

Peak shaving and load leveling are applications of demand-side management, which can benefit energy consumers, suppliers, and even housing construction companies.

- Energy consumers benefit in various ways. Reducing peak-time usage of grid-supplied electricity
  - saves energy costs as the battery is charged at night, when grid tariffs are lower, and supplies power during daytime peak hours;
  - complies with peak-time electricity usage rules, such as that set by the Ministry of Knowledge Economy of the Republic of Korea, which mandates a 10% reduction in peak-time electricity usage for the corporate sector;
  - reduces the use of electricity beyond the contracted amount, thus reducing overuse charges;
  - complies with carbon dioxide (CO_2) reduction regulations; and
  - prevents blackouts, by means of the UPS feature of generators.
- Energy suppliers save on peak power generation installation through an emergency power-saving subsidiary.
- Housing construction companies can attain a competitive edge when bidding for contracts that require CO_2 reduction and electric power saving.

3.6 MICROGRIDS

Microgrids (Figure 3.15) can be thought of as miniature electric power networks that can operate independently or while connected to the larger utility grid. Microgrids connected to an external grid are defined as interconnected loads and distributed energy resources within clearly defined electrical boundaries that act as a single controllable entity.

Microgrids have traditionally been used in industry for the purposes of power quality enhancement for critical loads and for cogeneration of heat and power. In recent times, microgrids using RESs have increasingly been used to power remote communities. However, a downside to inverter-based renewables is the lack of inertial response capability, which is the mechanical response of traditional synchronous generators to abrupt changes in system frequency. Community microgrids using inverter-based RESs are an important step toward energy security and sustainability.

Energy storage has many benefits for microgrids:

- providing ancillary services such as frequency regulation and voltage control, which are essential for microgrid operation;
- enhancing the integration of distributed and renewable energy sources;
- storing energy for use during peak demand hours;
- enabling grid modernization;
- integrating multiple smart-grid technologies;
- meeting end-user needs by ensuring energy supply for critical loads, controlling power quality and reliability at the local level; and
- promoting customer participation through demand-side management.

Figure 3.9 shows a typical microgrid installation on Jeju Island, Republic of Korea.

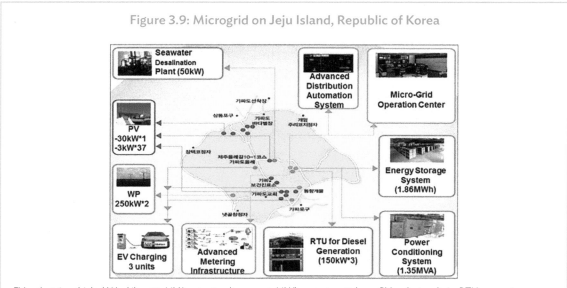

Figure 3.9: Microgrid on Jeju Island, Republic of Korea

EV = electric vehicle, kW = kilowatt, MVA = megavolt-ampere, MWh = megawatt-hour, PV = photovoltaic, RTU = remote terminal unit, Wp = watts of power.

Source: Korea Battery Industry Association 2017 "Energy storage system technology and business model".

4 CHALLENGES AND RISKS

4.1 GENERAL CHALLENGES

4.1.1 Cost Reduction

While many factors influence the growth of the ESS market, battery price is expected to have considerable impact on the viability of a BESS project. In recent years, the price of batteries used in BESSs has fallen rapidly. The price of lithium secondary batteries has dropped, from $1,000/kWh in 2010 to $227/kWh in 2016. The prices of other batteries are expected to fall by a further 50%–60% until 2030 (Figure 4.1).

According to the distribution of ESS system development installation costs, the battery cell accounts for 35%; power equipment, such as the BMS and the power conversion system (PCS), for 35%; and the construction cost of distribution and communication facilities, for 30%. As battery price makes up a large portion of the project cost, the current price falloff provides an opportunity for mainstream acceptance and use.

Battery prices are expected to drop even further in the future, and economies of scale can be realized through independent technology, investment in research and development (R&D), and expansion of productive capacity.

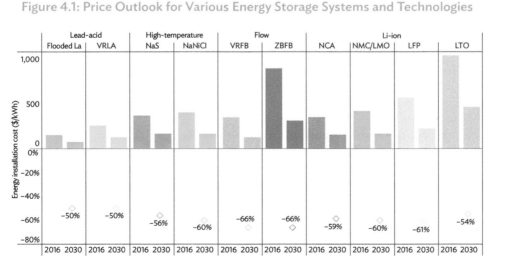

Figure 4.1: Price Outlook for Various Energy Storage Systems and Technologies

LA = lead-acid, LFP = lithium–iron–phosphate, LMO = lithium–manganese oxide, LTO = lithium–titanate, Na–NiCl$_2$ = sodium–nickel chloride, Na–S = sodium–sulfur, NCA = nickel–cobalt–aluminum oxide, NMC = nickel–manganese–cobalt, VRFB = vanadium redox flow battery, VRLA = valve-regulated lead-acid, ZBFB = zinc-bromine flow battery.

Source: IRENA (2017). Electric Storage and Renewable : Cost and Markets to 2030, IRENA, 2017

4.1.2 Deployment

ESSs can be used as power generation resources, in connection with the transmission and distribution network or with renewable energy, or as demand-side resources.

Use as power generation resource. This refers to the use of the ESS as power supply resource, which is the main role of power generators in existing power systems. In this role, the ESS can be used for power supply transfer (under a contract for difference) and for power supply capacity reinforcement.

Use in connection with transmission and distribution network. Unlike the use as power generation resource, connection to the T&D network allows the ESS to support the T&D facilities and provide temporary relief from distribution problems rather than supplying power continuously over an extended period. Like the use as power generation resource, on the other hand, ESS use in connection with the T&D network enables the supply or storage of the required power at the desired time through the charge and discharge capabilities of the ESS.

Use in connection with renewable energy sources. Electricity generation from RESs is greatly affected by natural conditions leading to variability and intermittence in the amount of power generated and the power generation period. Large deviations in solar and wind-power production from forecast volumes can put the stability of the power system at risk. An ESS can stabilize the power supply by storing power when demand or forecast error is low, and releasing it when power demand or forecast error is high.

Use as demand-side resource (for consumers). The use of the ESS for consumers is not very different from other uses mentioned above. The main difference among these various uses lies only in the beneficiary. Using the ESS may reduce the energy cost for consumers and improve the quality, service, and reliability of power (by providing a backup supply source).

4.1.3 Incentive Program

ESSs can help balance power demand and supply, thereby ensuring a stable power supply. Various demonstration projects and businesses centering on ESSs have been carried out globally for purposes such as reducing greenhouse-gas emissions and supporting aging power facilities.

The ESS market is still in its early stages, but it has been growing rapidly, mainly in Australia, Europe, Japan, and the United States (US). Key factors behind this growth are the fall in battery prices, improved stability of power systems, integration of alternative and renewable energy sources, and ESS policy. These elements are also expected to influence the increase in demand for ESSs for assisting power networks as power transmission and distribution costs rise with the aging of power networks.

At present, ESSs are still expensive, and the private sector is unlikely to take on large projects without a very clear business case on account of the high cost of initial installation and continued investment in battery cell replacement. However, some countries are making efforts to develop and supply ESS technology to the global market by providing subsidies for ESS installation and providing exemptions from related taxes. In some cases, governments are encouraging the private sector to participate in the market by providing subsidies to ease the burden of initial installation on consumers.

Regarding the subsidy policy of different countries, Germany is shouldering 30% of the installation costs of solar power generation–related ESSs with an ESS installation subsidy. The scale of German government support has increased, from €25 million in 2013 to €30 million between 2016 and 2018, for a total of about €150 million over 6 years. Japan also provided 31 billion yen in subsidies until 2015 in order to develop the industrial ecosystem for ESSs with the goal of capturing 50% of the global market by 2020. The US is urging power-market operators in each state to develop ESS-related business models, and the Republic of Korea is supporting ESS market penetration by extending the application of weighted value to new and renewable energy–related ESSs such as Photovoltaic 5.0 until 2019.

4.1.4　United Nations Framework Convention on Climate Change

The Paris Agreement, the new climate-change regime adopted by the United Nations Framework Convention on Climate Change (UNFCCC) in 2015, governs the climate change response from 2020 onward.[11] It will replace the Kyoto Protocol, which expires in 2020. While acknowledging the Intended Nationally Determined Contributions (INDCs) already submitted by the different countries, the agreement requires the submission of a higher goal every 5 years starting in 2020.

Developed countries have decided to provide at least $100 billion (about 118 trillion won) in support every year from 2020 onward to help developing countries cope with climate change. Unlike the Kyoto Protocol, which imposed the reduction obligation only on developed countries, this agreement is the first to require compliance by all 195 countries that ratified the UNFCCC.

To implement their commitments to reduce greenhouse gas emissions, the countries could introduce new and renewable energy in the power area and increase energy efficiency by promoting the wider-scale adoption of energy storage devices, to supplement variable new and renewable energy sources.

The Asian Development Bank (ADB) has raised $600 million to help finance climate change mitigation and adaptation projects with the issue of a 7-year green bond. In July 2017, ADB adopted its Climate Change Operational Framework 2017–2030. The framework strengthens ADB's support for compliance by its member countries with their climate commitments under the Paris Agreement, the Sustainable Development Goals, and the Sendai Framework for Disaster Risk Reduction 2015–2030, including their INDC pledges to reduce greenhouse-gas emissions.

ADB's climate mitigation and adaptation financing reached a record $4.5 billion in 2017, a 21% increase over the previous year's figure. ADB is now in a position to achieve its $6 billion annual climate financing target by 2020. Of this amount, $4 billion will be dedicated to mitigation through scaled-up support for renewable energy, energy efficiency, sustainable transport, and the building of smart cities, while $2 billion will be for adaptation through more resilient infrastructure, climate-smart agriculture, and better preparedness for climate-related disasters.

---

[11]　The 21st Conference of the Parties (COP21) of the UNFCCC held in Paris, France, from 30 November 2015 (local time), finally adopted the Paris Agreement on 12 December, after 2 weeks of negotiation.

4.2 GENERAL RISKS

4.2.1 Poorly Defined and Categorized Systems

ESS technologies are being actively developed and demonstrated around the globe. The ESS industry is receiving a tremendous boost from the increased integration of RESs into electric power systems. As the industry expands, defining and standardizing the terminology will become increasingly important.

Different terms are used in relation to ESSs in different countries and different fields of application (e.g., "ESS," "EES," "BESS"). Japan uses the term "electrical storage systems" in its technology standards and guidelines for electrical equipment to refer to electromechanical devices that store electricity. In the case of the US, the equivalent term is "rechargeable energy storage systems," defined in its National Electrical Code (NEC). In the Republic of Korea, while the term "electrical storage systems" can be found in its electrical equipment technology standards, what is mentioned in the standards established by the International Electrotechnical Commission (IEC) and adopted by the country are "electrical energy storage systems." The IEC standards define unit parameters and testing methods, and deal with planning, installation, safety, and environmental issues.

Setting standards to unify the unclear and vague terminology will improve technical communication between countries and institutions.

4.2.2 Unbundling of Operation and Network Development Activities

Unbundling rules restrict the ability of network operators (TSOs and distribution system operators) to engage in activities other than network operation. Such restrictions apply in particular to the management of power generation assets. The applicability of unbundling rules to EES is unclear at present and will depend on how storage is defined and categorized. In Europe, this matter needs to be decided primarily at the European Union (EU) level to ensure uniform application in all member states. Until this matter is clarified, regulatory uncertainty will remain. Classifying EES as a generation asset would make it more difficult for a network operator to have control over a storage project.

At present, there are differing views as to the necessity of applying the unbundling regime to EES. Since network operators could be important stakeholders in EES project development, and EES may contribute to resolving balancing issues cost-efficiently, one could argue that the regulatory framework should allow the participation of network operators in EES activities. But appropriate regulatory safeguards must be introduced to avoid undue distortions in competition resulting from the monopolistic position of network operators.

4.2.3 Grid Tariff Applications and Licensing Issues

ESSs store generated electricity using lithium-ion batteries and release the stored electricity when needed. Because of their high effectiveness in expanding the use of renewable energy and improving the efficiency of the power industry, ESSs can have a wide range of uses to power systems (suppliers) and users (consumers). The variability of renewable energy generation depending on climate conditions gives rise to the issue of concentrated power generation. The ability to store and manage electric power makes it possible to resolve variable generation and output concentration issues to a considerable degree. It also enables output regulation according to demand, thus reducing power consumption deviation (peak demand burden) per time period. ESS is therefore used to lessen problems with efficiency such as excess generation and excess facilities. Accordingly, power systems can use ESS to adjust frequency and to stabilize the output of renewable energy generators; users can use it to reduce peak power and lower fees or to sell surplus power according to policies.

In the Republic of Korea, an ESS powerhouse, policies are in place to procure a certain level of profits when installing and operating ESSs to reduce peak demand or when linking with solar power or wind power generation by the private sector to expand ESS distribution and improve the efficiency of the renewable energy and electric power industry. Profits from power generation comprise power sales profits and renewable energy certificate (REC) profits. When ESS is installed in wind power and solar power generation, and even if the same amount of energy is generated, support is offered to boost REC profits 4.5–5.0 times higher. Therefore, small-scale investments are possible, and demand to combine ESS mainly with solar power generation, which has high support benefits, continues to grow. The basic concept of peak demand reduction is difference transaction—using the difference in electricity fees per time period. Even if the same amount of electricity is used during the day, the peak load drops, thus lowering the base fee and the average cost of electricity. As ESS special fees will be applied until 2020, installing 1 MWh ESS for industrial use can save directly about 100 million won in electricity fees. If industrial electric fees are raised in the future, ESS demand for reducing peak demand is likely to grow mainly among manufacturers with high power usage.

4.2.4 Battery Safety

Among secondary batteries, lithium secondary batteries are the power source for the age in which mobile electronic devices have become commonplace. With application being expanded to industrial sectors such as battery electric vehicles (BEVs) and ESSs, R&D is being conducted on measures to ensure safety, such as lowering heating and ignition incidents resulting from the high density of lithium secondary batteries.

Lithium secondary batteries contain both oxidizers (negative) and fuel (positive) within the enclosed battery space, and therefore also carry the risk of fire and explosion in case of overcharging, over-discharging, excess current, or short circuits (Figure 4.2). For battery safety, safety design is essential at the cell, module, pack, and final product level. If safety fails at one level, more severe accidents at the higher levels can quickly follow. There is no single standard or parameter for assessing battery safety. A battery protection circuit will improve safety by making such accidents less likely or by minimizing their severity when they do occur.

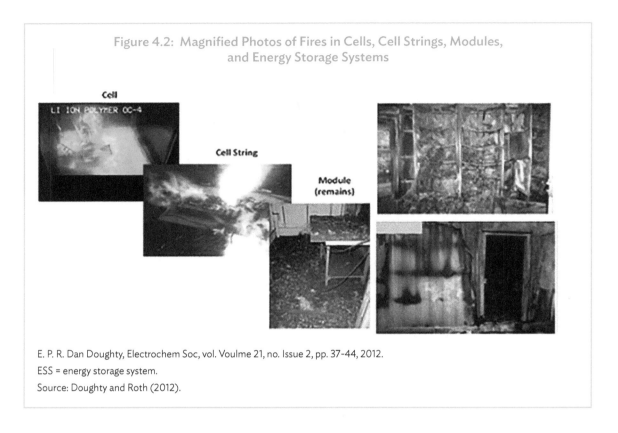

Figure 4.2: Magnified Photos of Fires in Cells, Cell Strings, Modules, and Energy Storage Systems

E. P. R. Dan Doughty, Electrochem Soc, vol. Voulme 21, no. Issue 2, pp. 37-44, 2012.

ESS = energy storage system.

Source: Doughty and Roth (2012).

Protection devices are integrated into the cell, module, and battery systems to prevent abnormalities and cut down on accidents. Current interrupt devices (CIDs), positive temperature coefficient (PTC) thermistors, current-limiting fuses, diodes, battery management systems (BMSs), etc., control the occurrence and intensity of heat and gas. Moreover, the need for fire suppression systems is getting increasing attention as plans are made to reduce the severity of accidents.

However, there are currently no standards or test criteria for fire prevention systems design. In the US, testing of such systems is in its initial phases, and the Fire Protection Research Foundation, an affiliate of the nonprofit National Fire Protection Association (NFPA), is winding up research on fire suppression, including risk management of lithium secondary batteries and sprinkler design.

4.3   CHALLENGES OF REDUCING CARBON EMISSIONS

The impact on CO_2 emissions of electrical storage systems in power systems with high penetrations of wind generation, such as the Irish power system, can be used as an example. The observed dispatch of each large generator in 2008–2012 was used to estimate a marginal emission factor of 0.547 kgCO$_2$/kWh (McKenna, Barton, and Thomson 2017).

Selected storage operation scenarios were used to estimate storage emission factors—the carbon emission impact associated with each unit of storage energy used. This value is substantially higher than the estimated average emission factor for the same period (0.489 kgCO$_2$/kWh), highlighting the potential to underestimate the impact of demand-side interventions if the lower value is used incorrectly. With the aim of estimating the short-run in-use environmental impact of electricity storage in the Irish power system, marginal emission data were filtered according to various storage operation scenarios to estimate marginal emission factors for storage charging and discharging. These were combined with storage round-trip efficiency to provide an estimate of the "storage emission factor"—the carbon emission impact associated with each unit of energy delivered from storage.

$$CO_2 \; reduction \; (TOE) \; = \; ESS \; capacity \; [MW] \; * \; usage \; \left[\frac{h}{y}\right] * useful \; life \; [y] \; * \; 0.5642 \; \left[\frac{TOE}{MWh}\right]$$

$$e.g. \; 500 \; MW \; * \; 1\frac{h}{d} \; * \; 365\frac{d}{y} \; * \; 10 \; y \; * \; 0.5642 \; \frac{TOE}{MWh} \; = \; 1{,}029{,}665 \; TOE$$

CO2 = carbon dioxide, d = day, ESS = energy storage system, h = hour, MW = megawatt, MWh = megawatt-hour, TOE = metric tons of oil equivalent, y = year.

Table 4.1 lists the various energy storage applications together with their emission reduction potential.

Table 4.1: Energy Storage Services and Emission Reduction

Category	Service	Potential for Emission Reduction	Nature of Direct Emission Reduction
Bulk energy services	Electric energy time shift (arbitrage)	Yes	If energy storage avoids or reduces curtailment, additional emission reductions can be made to the extent that the additional renewables displace higher emission sources of generation. Where a renewable generator is able to shift grid output from one time period to another, higher or lower emission reduction may result, depending on the difference between the grid-emission factors at the time of dispatch.
	Electric supply capacity	Yes	To the extent that energy storage technologies allow higher levels of renewable-energy generation, additional emission reductions are possible.
Ancillary services	Frequency regulation	Maybe	Where energy storage technologies can replace frequency regulation services traditionally provided by fossil fuel generators, there may be potential for reducing emissions. This depends on the source of the power used to charge the energy storage system, which is then used periodically for frequency regulation services.
	Spinning, non-spinning, and supplemental reserves	Maybe	As above, potential for emission reduction depends on the source of the power used to charge the energy storage system.
	Voltage support	Maybe	As above
	Black start	Maybe	As above
Transmission infrastructure services	Transmission upgrade deferral	Maybe	Depends on analysis of various options and source of power used to charge the energy storage system
	Transmission congestion relief	Yes	Where energy storage technologies can help avoid the curtailment of renewable energy sources due to transmission congestion, there is potential for contribution to emission reduction.

continued on next page

Table 4.1 *continued*

Category	Service	Potential for Emission Reduction	Nature of Direct Emission Reduction
Distribution infrastructure services	Distribution upgrade deferral	Maybe	Depends on analysis of various options and source of power used to charge energy storage system
	Voltage support	Maybe	As above. Potential for emission reduction depends on the source of the power used to charge the energy storage system.
Customer energy management services (residential, and C&I)	Power quality	Maybe	As above
	Power reliability	Maybe	As above
	Retail electric energy time shift	Maybe	Time shifting—using the energy storage system when grid emission intensity is low and discharging when grid emission intensity is higher—offers potential for additional emission reduction.
	Demand charge management	Maybe	Demand charge management is a form of time shifting and so, as above, the potential for emission reduction depends on the source of the power used to charge the energy storage system.
	Increased self-consumption of solar PV	Maybe	If the availability of energy storage leads households to install larger solar PV systems, there could be additional emission reduction. Otherwise, time shifting by itself will not in itself reduce emissions.
Off-grid	Solar home system	Yes	Where off-grid energy storage systems replace biomass or fossil fuel alternatives, emission reduction will be possible.
	Mini-grids: System stability services	Yes	As above
	Mini-grids: Facilitation of high share of variable renewable energy	Yes	As above
Transport sector	Transport applications of energy storage are not a focus of this program		

C&I = commercial and industrial, PV = photovoltaic.

Source: ADB, based on IRENA taxonomy of energy storage services.

4.4 BATTERY RECYCLING AND REUSE RISKS

Battery reuse (Figure 34) is defined under Article 3.13 of the EU's Waste Framework Directive 2008/98/EC as any operation that calls for using batteries again for the same purpose for which they were originally designed. One could argue that any reuse of batteries—a system component— must occur in the original application in order for all technical performance and safety aspects to be secured. This is also indicated in the End-of-Life Vehicles Directive 2000/53/EC, Article 2.6, of the EU, which defines reuse as any operation in which components of end-of life vehicles are used for the same purpose for which they were conceived.

Battery recycling is aimed at reducing the number of batteries disposed of as municipal solid waste. It is the best approach to end-of-life management of spent batteries, mainly for environmental, but also for resource conservation and economic reasons.

Battery recycling plants require the sorting of batteries according to their chemistry. Some sorting must be done before the batteries arrive at the recycling plant. Nickel–cadmium, nickel–metal hydride, lithium-ion, and lead–acid batteries are placed in designated boxes at the collection point. Battery recyclers claim that if a steady stream of batteries, sorted by chemistry, were made available at no charge, recycling would be profitable. But preparation and transportation add to the cost.

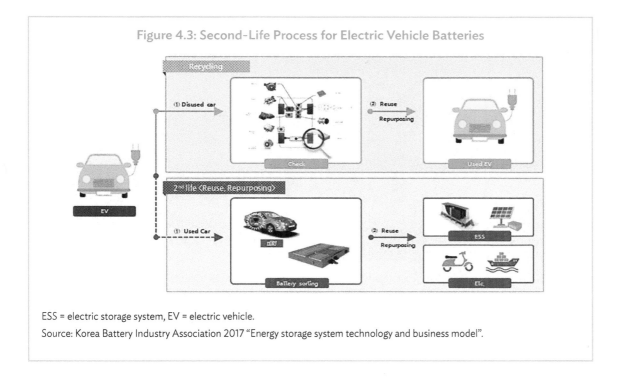

Figure 4.3: Second-Life Process for Electric Vehicle Batteries

ESS = electric storage system, EV = electric vehicle.

Source: Korea Battery Industry Association 2017 "Energy storage system technology and business model".

4.4.3   Examples of Battery Reuse and Recycling

General Motors used-battery electric storage system project with ABB. Used Chevrolet Volt batteries are helping keep the lights on at the new General Motors (GM) Enterprise Data Center at GM's Milford Proving Ground in the US state of Michigan.

Figure 4.4 shows a microgrid backup system powered by five used Chevrolet Volt batteries—the result of collaboration between GM and Swiss power engineering firm ABB.

BMW used-battery electric storage system project with Bosch. Electro-mobility and power storage are two core elements of the move to alternative forms of energy. A project is bringing the German-based multinational engineering and electronics company Bosch together with the BMW Group and Swedish power company Vattenfall to drive progress on both technologies by interconnecting used batteries from electric vehicles to form a large-scale ESS in Hamburg. The energy produced, available within seconds, can help keep the power grid stable.

Figure 4.4: GM–ABB Second-Life Electric Vehicle Battery Applications

Source: Charged Electric Vehicles Magazine, "Nissan, GM and Toyota repurpose used EV batteries for stationary storage," Gharged Electric Vehicles Magazine, 17 6 2015. [Online]. Available: https://chargedevs.com/newswire/nissan-gm-and-toyota-repurpose-used-ev-batteries-for-stationary-storage/. Morris (2015).

Li-ion batteries still have high storage capacity at the end of their life cycle in electric vehicles. For this reason, they are still very valuable and can be used extremely efficiently as stationary buffer storage for many more years (Figure 4.5). The project allows the three partners to gain many new insights into potential areas of application for such batteries, their aging behavior, and their storage capacity. Bosch's management algorithm is applied to ensure maximum service life and performance, as well as other benefits (Figure 4.6).

Renault used-battery project with Powervault. Automakers like Renault have introduced a new business model within the framework of battery pack reusability. In this model, the battery pack is leased to the vehicle owner, while actual ownership

Figure 4.5: Second-Life Energy Storage Application for BMW Electric Vehicle Batteries

Source: Chris Davies. BMW's battery graveyard gives EV cast-offs a second life, October 2017. https://www.slashgear.com/bmws-battery-graveyard-gives-ev-cast-offs-a-second-life-26505613/

Figure 4.6: BMW–Bosch Second-Life Electric Vehicle Battery Demonstration Project

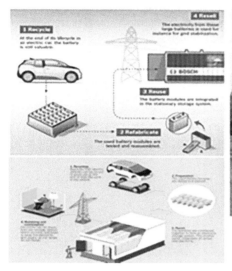

Source: M. KANE, "Bosch Cooperates With BMW And Vattenfall In Second Life Battery Project," Inside EVs, 9 2 2015. [Online]. Available: https://insideevs.com/bosch-cooperates-with-bmw-and-vattenfall-in-second-life-battery-project/. Kane (2015).

is retained by the manufacturer. When these battery packs reach the end of their operational life, the automaker replaces them with new battery packs at a fraction of the cost of the actual battery. Two new energy storage units made by Connected Energy of the UK use old Renault electric vehicle batteries, and were recently installed at fast-charging stations on highways in Belgium and Germany.

Renault and London-based Powervault, manufacturer of home ESSs, have announced a partnership to promote the reuse of Renault electric vehicle batteries for home energy storage units, thus reducing the cost of a Powervault smart battery unit by 30% (Figure 4.7). The partnership kicked off with a year-long trial of 50 units installed in solar-powered homes in the UK. During the trial, which began in July 2017, system performance was monitored and customer reaction ascertained. The results of the trial will help in the preparation of a rollout strategy for general release.

Figure 4.7: Renault–Powervault's Second-Life Electric Vehicle Battery Application

Source: M. KANE, "Renault To Enter Home Battery Market With Repurposed EV Batteries," InsedEVs, 26 6 2017. [Online]. Available: https://insideevs.com/renault-repurposed-ev-batteries-ess/. Kane (2017).

Nissan used-battery ESS project with 4R Energy. A joint venture between Sumitomo and Nissan called 4R Energy—"4R" stands for "Reuse, Resell, Refabricate, and Recycle"—uses 16 lithium-ion batteries from electric vehicles to help monitor energy fluctuations and store the solar farm's energy output (Figure 4.8).

Figure 4.8: Nissan–Sumitomo Electric Vehicle Battery Reuse Application (4R Energy)

Source: Jim (2014). Jim, "Used Nissan EV Batteries Now Provide Grid Scale Storage," Vehicle to Grid UK, 11 5 2014. [Online]. Available: http://www.v2g.co.uk/2014/05/used-nissan-ev-batteries-now-provide-grid-scale-storage/.

The human-made island of Yumeshima in Osaka, western Japan, is now home to the world's first large-scale energy storage system, highlighting the potential for reusing electric vehicle batteries.

4.4.1   Reuse of Electric Vehicle Batteries for Energy Storage

The end-of-life (EOL) of a battery is around 80% of initial capacity. However, even at 80% capacity, the battery can be used for 5–10 more years in ESSs (Figures 4.9 and 4.10).

Figure 4.9: Reuse of Electric Vehicle Batteries In Energy Storage Systems

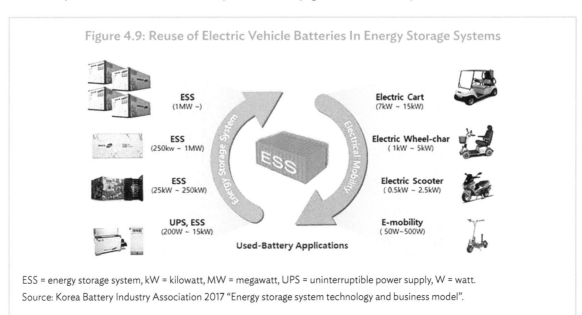

ESS = energy storage system, kW = kilowatt, MW = megawatt, UPS = uninterruptible power supply, W = watt.

Source: Korea Battery Industry Association 2017 "Energy storage system technology and business model".

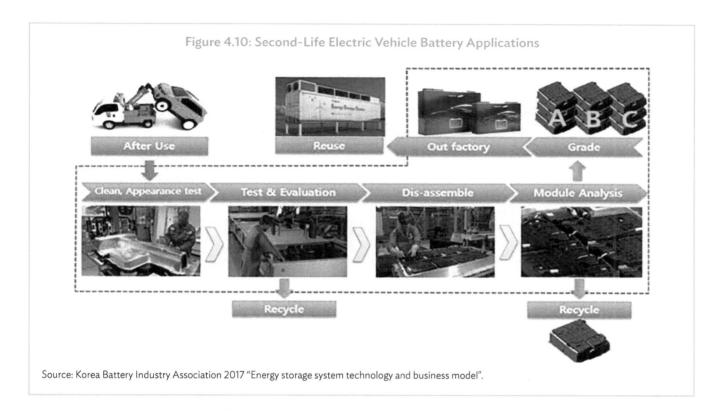

Figure 4.10: Second-Life Electric Vehicle Battery Applications

Source: Korea Battery Industry Association 2017 "Energy storage system technology and business model".

4.4.2 Recycling Process

The recycling process starts by removing combustible materials, such as plastic insulation, with a gas-fired thermal oxidizer. Gases from the thermal oxidizer are sent to the plant's scrubber, where they are neutralized to remove pollutants. The process leaves the clean, naked cells, which contain valuable metal content. The cells are then chopped into small pieces, which are heated until the metal liquefies. Non-metallic substances are burned off, leaving a black slag on top that is removed with a slag arm. The different alloys settle according to their weight and are skimmed off like cream from raw milk. Li-ion batteries are currently reprocessed through pyrolysis (heat treatment), with the metal content as primary recovery. Zinc–carbon and/or air and alkaline–manganese batteries can be reprocessed using different methods, including smelting and other thermal–metallurgical processes to recover the metal content (particularly zinc). The processing method includes purification and separation through hydrometallurgy processing after physical pretreatment. Hydrometallurgical processing involves dissolving the spent lithium-ion batteries and then selectively separating them from the leach liquor, which is then purified to obtain the required valuable metals. Through crushing and shredding, the usual pretreatment operations in hydrometallurgical processing, the materials are easily liberated. Most processing plants therefore use a combination of hydrometallurgical and mechanical processing (Figure 4.11).

Figure 4.11: Lithium-Ion Battery Recycling Process

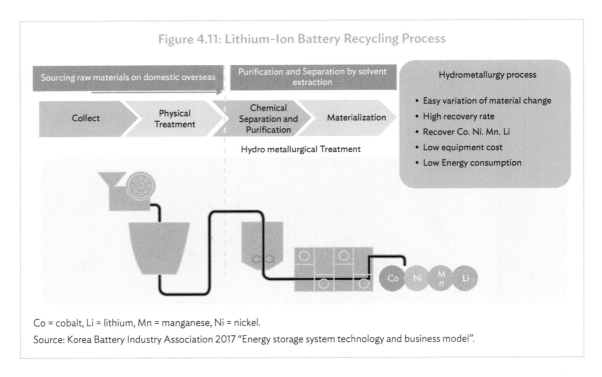

Co = cobalt, Li = lithium, Mn = manganese, Ni = nickel.
Source: Korea Battery Industry Association 2017 "Energy storage system technology and business model".

Chemical treatment process (hydro-process). The chemical recycling process applied to lithium batteries and the materials resulting from the process are shown in Figure 4.12.

Physical separation and purification of cathode active material. Discharged Li-ion batteries are placed in an inert, dry atmosphere for mechanical crushing and shredding. This reduces the impact of internal short circuits when in contact with oxygen, and avoids exposing the materials to water vapor, which would hydrolyze the electrolyte.

Figure 4.12: Chemical Recycling of Lithium Batteries, and the Resulting Materials

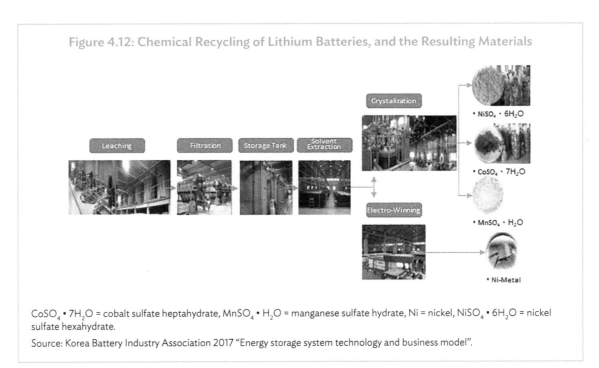

$CoSO_4 \cdot 7H_2O$ = cobalt sulfate heptahydrate, $MnSO_4 \cdot H_2O$ = manganese sulfate hydrate, Ni = nickel, $NiSO_4 \cdot 6H_2O$ = nickel sulfate hexahydrate.
Source: Korea Battery Industry Association 2017 "Energy storage system technology and business model".

Plastics packaging is set apart for recycling, while electrodes and electrolytes undergo further processing (Figure 4.13).

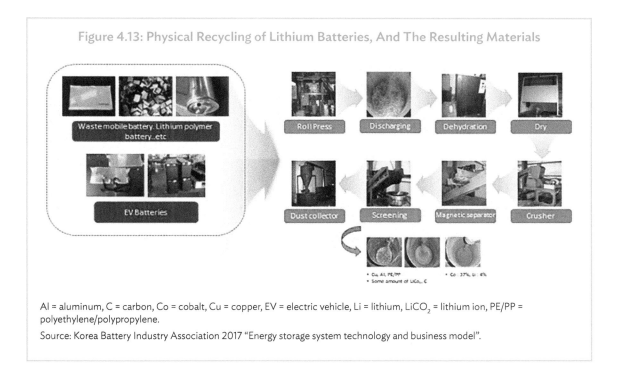

Figure 4.13: Physical Recycling of Lithium Batteries, And The Resulting Materials

Al = aluminum, C = carbon, Co = cobalt, Cu = copper, EV = electric vehicle, Li = lithium, LiCO$_2$ = lithium ion, PE/PP = polyethylene/polypropylene.

Source: Korea Battery Industry Association 2017 "Energy storage system technology and business model".

5 POLICY RECOMMENDATIONS

The business case for the installation of both large-scale and small-scale battery energy storage requires supportive energy policies. Recommendations for select use cases are provided below.

5.1 FREQUENCY REGULATION

Responsibility for frequency regulation is almost always assigned to the TSO of the synchronous control area. BESSs have proven to be technically capable of delivering frequency regulation. In fact, the response time of a BESS (in sub-seconds) is much faster than that of a conventional power plant (typically 3–5 seconds). Therefore, energy policy should reflect the technical capabilities of various asset types including BESS for use in frequency regulation.

Policy recommendations. The policy recommendations pertaining to frequency regulation are as follows:

Update the grid code to allow for the delivery of frequency regulation by means of various generation, load, and storage configurations.

Create market instruments that capitalize on the fast response time of batteries. For example, the National Grid ESO has created a new service known as enhanced frequency response (EFR), which requires faster ramp rates and response times to reflect the enhanced capabilities of BESS in frequency regulation. In Germany, the regulator has relaxed the bidding criteria to increase the participation of BESS and other asset types in existing ancillary service markets.

Enable smaller participants such as BESS operators and "prosumers" (producers–consumers) to offer ancillary services by reducing the minimum bid size (MW) and volume of available energy (MWh). For example, the German regulator has changed secondary reserve procurement from weekly to daily delivery.

5.2 RENEWABLE INTEGRATION

5.2.1 Distribution Grids

The major challenge presented by grid integration of renewable energy sources is cost recovery of network investment. In most countries, the regulatory framework incentivizes network operators to size networks according to peak capacity. In some countries where renewable energy sources are guaranteed grid access, curtailment of renewable energy due to insufficient grid capacity can lead to fines and penalties for the network operator. However, cost recovery of network investments is achieved mostly, if not entirely, from consumers through use-of-system charges. The burden of network expansion related to grid-integration of RES therefore falls on consumers.

Another significant challenge that can be addressed with BESS is managing power quality in distribution networks. Traditionally, the feeder voltage in the substation would be kept close to the upper voltage tolerance limit to enable sufficient voltage drops (as power is consumed) for all consumers along the length of the feeder. However, with the rise of the prosumer, network operators must be vigilant against increases in voltage during peak solar hours or on weekends, when solar supply exceeds electricity consumption.

Distribution grid operations can be improved by means of a BESS:

Network congestion. In locations where network capacity is limited during peak renewable-generation hours, a BESS can store the excess energy and release it into the network when renewable generation reduces.

Power quality. A BESS can be used to absorb the supply of renewable energy and keep the voltage below the upper limit prescribed in the grid code. The BESS can be either a grid-tied or a behind-the-meter installation.

5.2.2 Transmission Grids

For transmission network operators, a key concern arising from the integration of renewable energy sources is the effect of the variability and intermittence of generation. The various problems created, such as the following, can be addressed with the help of BESS:

Forecast errors. Some degree of forecast error is always likely because of the variability of renewable energy sources. To mitigate the risk of a large unexpected deviation from the forecast production of renewable energy, the TSO is forced to increase its operational reserves. In such cases, a BESS can reduce the forecast error by smoothing the energy entering the electricity network, thereby reducing the forecast error.

Network congestion. High volumes of RES generation can sometimes lead to network congestion. The TSO is then forced to re-dispatch—change the dispatch level of some power plants from the least-cost result of the unit commitment algorithm—because of network constraints. The only other option would be to curtail RES generation. The increased cost burden in both instances falls on the consumer. In such cases, a BESS can store energy during periods of network congestion, thereby reducing the need for RES curtailment or power plant re-dispatch.

Increased ramping requirements during evening peak hours. To deal with the increase in ramping requirements, a BESS can store energy during a period of high renewable-energy production, e.g. solar peak hours, and release it into the grid during off-peak production hours, e.g., during the evening hours, when solar production drops off and the evening peak demand increases.

Policy recommendations. The following recommendations are aimed at dealing with the problems of grid transmission:

Create regulatory frameworks that encourage the use of ESSs to defer network expansion and reduce the associated consumer burden. The cost of the energy storage should instead be borne by the RES asset owner.

Impose stiff penalties on utility-scale grid-connected RES owners when energy production differs greatly from forecasts. This measure will incentivize RES owners to install energy storage systems to reduce the forecast error.

Require distribution grid–connected RES and prosumers to install BESS to increase self-consumption or delay supply until notified in locations where voltage rise is a problem, or risk being curtailed without recourse.

Incentivize energy storage systems via a capacity market or other market mechanisms to replace expensive "peaker" power plants during the evening ramp hours.

5.3   PEAK SHAVING AND LOAD LEVELING

Network operators should be required to identify sites where it makes sense to use storage for peak shaving instead of conventional network reinforcement. These projects should then be put up for competitive bidding to storage developers to ensure low costs for consumers.

5.4   MICROGRIDS

Microgrids that use renewable energy sources for power generation typically have a diesel generator set (genset) to provide residual power and ancillary services. In such cases, the use of an energy storage system to provide backup power and ancillary services should be encouraged in place of the diesel genset.

A-1. SAMPLE FINANCIAL AND ECONOMIC ANALYSIS

Table A.1: Underlying Assumptions

Item/Metric	Definition/Value
Business model	Frequency regulation
BESS specification	20 MW/5 MWh
Daily battery cycling	8 cycles
Depth of discharge (DOD)	80%
Round trip efficiency	85%
Daily energy use	37.65 MWh*
Useful life (battery cells)	15,000 cycles / 8 cycles per day => 5 years
Useful life (balance of plant)	10 years
Total project life	10 years
Debt/Equity	80%/20%
Cost of debt	10%
Cost of equity	15%
WACC	11%
Tax rate	30%

BESS = battery energy storage system, MW = megawatt, MWh = megawatt-hour, WACC = weighted average cost of capital.

*Daily energy use = BESS power (20 MW) * capacity (5 MWh) * round trips per day (8 cycles) * DOD per round-trip (80%)/round trip efficiency (85%) = 37.65 MWh.

Source: Korea Battery Industry Association 2017 "Energy storage system technology and business model".

Table A.2: Capital Expenditure

Component	Content	Component Cost mio. USD	Units	Total Cost mio. USD
Battery	5 MWh * USD 0.4 million per MWh	2		2
Power control system (PCS)	20 MW * USD 0.15 million per MW	3		3
HV Transformer	220 kV / 33 kV, 30 MVA	0.5	1	0.5
MV Transformer	33 kV / 0.44 kV, 5 MVA	0.08	6	0.48
Storage Containers for Battery		0.05	6	0.3
Storage Containers for PCS		0.05	6	0.3
Installation		0.1		0.1
Battery cell replacement (Year 5)		0.5		0.5
TOTAL PROJECT COST				7.18

HV = high voltage, kV = kilovolt, MV = medium voltage, MVA = megavolt-ampere.

Notes:

Based on average daily use, it is estimated that the battery cells will have to be replaced in year 5.
The cost of replacement cells in year 5 is assumed to be 50% of current prices.
The salvage price of the original cells in year 5 is assumed to be 25% of current prices.

Source: Korea Battery Industry Association 2017 "Energy storage system technology and business model".

Table A.3: Operating Expenditure

Energy Cost	
Daily energy charge	37.65 MWh
Annual energy charge	13,741 MWh
Power purchase price	$80/MWh
Annual power purchase cost	$1.1 million

Table A.4: Revenue

Energy Revenue*	GBP/MW	$/MW
EFR market clearing price	11.97	16.725
Annual revenue per MW		$146,511
Annual revenue (20 MW)		$2.93 million

EFR = enhanced frequency response, GBP = pounds sterling, MW = megawatt.

*Based on 2016 auction for EFR batteries by the National Grid (UK transmission system operator).

Table A.5: Financial Internal Rate of Return

Item	Value
Project IRR	9.76%
Project NPV	($320,745.74)

IRR = internal rate of return, NPV = net present value.

Table A.6: Calculation of Financial Internal Rate of Return

Year	0	1	2	3	4	5	6	7	8	9	10
EFR Revenues		2,930,220	2,930,220	2,930,220	2,930,220	2,930,220	2,930,220	2,930,220	2,930,220	2,930,220	2,930,220
Energy purchased		(1,099,294)	(1,099,294)	(1,099,294)	(1,099,294)	(1,099,294)	(1,099,294)	(1,099,294)	(1,099,294)	(1,099,294)	(1,099,294)
Insurance (1% of capital cost per year)		(71,800)	(71,800)	(71,800)	(71,800)	(71,800)	(71,800)	(71,800)	(71,800)	(71,800)	(71,800)
O&M (4% of EPC contract)		(28,720)	(28,720)	(28,720)	(28,720)	(28,720)	(28,720)	(28,720)	(28,720)	(28,720)	(28,720)
EBITDA		1,730,406	1,730,406	1,730,406	1,730,406	1,730,406	1,730,406	1,730,406	1,730,406	1,730,406	1,730,406
Less: D&A (20% p.a. straight-line)		(1,436,000)	(1,436,000)	(1,436,000)	(1,436,000)	(1,436,000)	(100,000)	(100,000)	(100,000)	(100,000)	(100,000)
EBIT		294,406	294,406	294,406	294,406	294,406	1,630,406	1,630,406	1,630,406	1,630,406	1,630,406
Less: Interest Expense		(574,400)	(538,359)	(498,714)	(455,104)	(407,134)	(354,366)	(296,322)	(232,473)	(162,240)	(84,983)
Less: Taxes		83,998	73,186	61,292	48,210	33,818	(382,812)	(400,325)	(419,380)	(440,450)	(463,677)
Tax Net Income		(195,996)	(170,767)	(143,016)	(112,489)	(78,910)	893,228	933,859	978,553	1,027,716	1,081,796
Add: D&A		1,436,000	1,436,000	1,436,000	1,436,000	1,436,000	100,000	100,000	100,000	100,000	100,000
Capex	(7,180,000)					(500,000)					
After Tax Levered Cash Flow	(7,180,000)	1,240,004	1,265,233	1,292,984	1,323,511	857,090	993,228	1,033,859	1,078,553	1,127,716	1,181,796
Levered Project IRR	9.76%										
Levered Project NPV	($320,745.74)										

D&A = depreciation and amortization; EBIT = earnings before interest and taxes; EBITDA = earnings before interest, tax, depreciation, and amortization; EFR = enhanced frequency response; IRR = internal rate of return; NPV = net present value; O&M = operation and maintenance.

Source: Korea Battery Industry Association 2017 "Energy storage system technology and business model".

Note: To calculate the economic internal rate of return:

- Replace EFR earnings with economic benefit (incremental and non-incremental).
- Use 9% discount rate in NPV calculation.

For comparison, 100-megawatt-equivalent capacity storage of each resource type was considered. In the solar-plus-storage scenario, the following assumptions were made: 100-megawatt (MW), 3-hour lithium-ion battery energy storage system coupled with a 50 MW solar photovoltaic system, and a project life of 20 years.

Energy storage technology cost assumptions were selected by means of projected cost information collected from vendors and public information sources (University of Minnesota Energy Transition Lab, Strategen Consulting, and Vibrant Clean Energy 2017).[1]

With this information, the installed cost for a 4-hour, 100 MW lithium-ion battery storage system was assumed to be about $1,600 per kilowatt for a 2018 start date, representing the total all-in cost of the storage medium; the power conversion system; engineering, procurement, and construction (EPC); replacements; and other ongoing and recurring costs.

The table below shows that energy storage installed costs are foreseen to improve to $1,200/kW by 2023. Improvements are also anticipated in the fixed O&M costs and round trip efficiencies over time.

Table A.7: Calculation of Financial Internal Rate of Return
(University of Minnesota Energy Transition Lab, Strategen Consulting, and Vibrant Clean Energy 2017)

Scenario:	Storage Only (2018)	Storage Only (2023)	Solar + Storage (2018)	Solar + Storage (2023)
Size/Duration	100 MW/ 4 hrs	100 MW/ 4 hrs	100 MW/ 3 hrs	100 MW/ 3hrs
Installed Cost (4-hrs)	$1600/kW	$1200/kW	$1335/kW	$1020/kW
Fixed O&M	$16/kW-yr	$14/kW-yr	$16/kW-yr	$14/kW-yr
Variable O&M	$4/MWh	$4/MWh	$4/MWh	$4/MWh
Round Trip Efficiency (incl. auxiliaries)	85%	90%	85%	90%

Scenario:	Storage Only (2018)	Storage Only (2023)	Solar + Storage (2018)	Solar + Storage (2023)
Installed Cost	$829/kW	Base Case: $829/kW Sensitivity: $1200/kW	$829/kW	$829/kW
Fixed O&M	$8.50/kW-yr	$8.50/kW-yr	$8.50/kW-yr	$8.50/kW-yr
Variable O&M	$2.30/MWh	$2.30/MWh	$2.30/MWh	$2.30/MWh
Capacity Factor	10%	10%	10%	10%
Heat Rate	9,750 BTU/kWh	Base Case: 9,750 BTU/kWh Sensitivity: 9,300 BTU/kWh	9,750 BTU/kWh	9,750 BTU/kWh

BTU = British thermal unit, kW = kilowatt, kWh = kilowatt-hour, MWh = megawatt-hour, O&M = operation and maintenance.

Source: P. Sheilds, "Modernizing Minnesota's Grid : An Economic analysis of Energy storage opportunities," in Minnesota Energy Storage Strategy Workshop, July, 11, 2017.

---

1   P. Sheilds, "Modernizing Minnesota's Grid : An Economic analysis of Energy storage opportunities," in *Minnesota Energy Storage Strategy Workshop*, July, 11, 2017.

For solar plus storage, dispatch is optimized to ensure that 75% of charging energy comes from eligible renewable resources. This was applied to the portion of the project's storage equipment costs corresponding to the fraction of output energy that is charged directly from renewable resources (solar photovoltaic). For projects starting in 2023, the 22% applied assumes that projects begin construction before 31 December 2021.

B-1. CASE STUDY OF A WIND POWER PLUS ENERGY STORAGE SYSTEM PROJECT IN THE REPUBLIC OF KOREA

The installation price of 20 megawatts (MW) of wind power was set at 2,494 thousand won/kilowatt (kW), the rate provided by the Korea Energy Agency. For the average usage rate of wind-power generators, the average usage rate of 23.13% for a specific wind power complex in the Republic of Korea was applied.

Table B.1: Major Premises and Assumptions for Simple Levelized Cost of Electricity Estimations of Wind Power

Category	Item	Price (Unit Cost)	Remarks
Technological premise	Total installation capacity	20 MW	
	Installation price	2,494 thousand won/kW	Internal data of the Korea Energy Agency (2013)
	Fixed cost	2.5% of total construction cost	
	Variable cost	0.8% of total generation cost	On-site consumption rate
	Average usage rate	21.13%	Product of specific company in specific region
	Performance decline rate	0%	
Financial/ Economic premise	Facility life	20 years	
	Discount rate	5.5%	KDI public investment project discount rate
	Inflation rate	3%	
	Won/dollar exchange rate	1,095	Average exchange rate in 2013
ESS	Total facility capacity	2 MWh	10% of total wind-power capacity of facility
	Installation price	1.5 billion/MWh	KEPCO's frequency adjustment ESS installation project
	Lithium cell price	$500/kWh	BNEF
	Life	10 years	4,900 cycles (replace after 10 years)
	Loss rate	15%	Charging/discharging and standby loss

BNEF = Bloomberg New Energy Finance, ESS = energy storage system, KDI = Korea Development Institute, KEPCO = Korea Electric Power Corporation, kW = kilowatt, kWh = kilowatt-hour, LCOE = levelized cost of electricity, MW = megawatt, MWh = megawatt-hour.

The generation price of wind power plus energy storage system (ESS) is 167.4 won per kilowatt-hour (kWh), higher than that for gas turbine generators. When only wind power is installed, the generation price is 153.9 won. This figure is only slightly higher than the 2013 average system marginal price (SMP) of 152.1 won/kWh, showing that wind-power generation is close to grid parity.

Installing ESS raises the generation price above the SMP, thus reducing economic feasibility. Also, adding ESS to wind power increases the total construction cost by 6.0%, from 49,880 million won to 52,880 million won, and the generation price by 8.8%. The increase in generation price is higher than the rise in construction cost mainly because, while no additional facilities are installed, passage through the battery adds a 15% ESS charging and discharging and standby loss rate for the generated amount, thus lowering the amount of commercial power supply. When only 20 MW of wind power is installed, the annual generation amount expected in the first year is 40.5 gigawatt-hours (GWh); when ESS is added, the amount of generation that can be commercially supplied drops by 1.8%, to 39.5 GWh.

In the case of wind power, the power price (commercial levelized cost of electricity, or LCOE) must be at least 181.8 won/kWh—8.6% higher than the generation price (simple LCOE) of 167.4 won/kWh—for wind power plus ESS to be commercially feasible.

Table B.2: Comparison of Levelized Cost of Electricity for Wind Power Generation at Various Energy Storage System Operating Rates

ESS Operating Rate	Simple LCOE (won/kWh)	Commercial LCOE (won/kWh)	Total Investment Cost (billion won)	ESS Investment Cost (billion won)
0%	153.9	167.4	49.9	–
5%	160.6 (4.34%)	174.5 (4.27%)	51.4 (3.01%)	1.5
10%	167.4 (8.76%)	181.8 (8.62%)	52.9 (6.01%)	3.0
15%	174.3 (13.26%)	189.2 (13.05%)	54.4 (9.02%)	4.5

ESS = energy storage system, LCOE = levelized cost of electricity.
Note: Digits within parentheses show the rate of increase compared with the ESS capacity rate of 0%.

The relationship between the increase in generation price and the increase in total investment costs shows that when the ESS capacity rate rises from 0% to 5%, the total investment cost rises by only 3.01%; the generation price, on the other hand, goes up by 4.27%, a rate 1.44 times higher than the rate of increase in investment cost.

The difference between the two rates of increase becomes 1.46 times higher at an ESS operating rate of 10%, and 1.47 times higher at an ESS operating rate of 15%. The minimum power cost for commercial feasibility exhibits trends similar to that for generation cost.

Like solar power generation, wind power generation has several marked characteristics:

- When investments for wind power generation facilities increase, the generation cost does not change much, but when ESS is added to the generation facilities, the generation cost increases.

- When ESS investments increase, the amount of increase in generation costs also goes up.
- The increase rate for generation cost is about 1.4 times higher than the increase rate for investment scope.

An ESS charging and discharging and standby loss rate of 15% was assumed in the foregoing analysis, which shows the generation cost and the amount of generation in the first year of operation at an ESS charging rate of 15%, a discharging rate of 20%, and standby loss rate of 25%. Installing an ESS with an operating rate of 15% reduces the amount of generation by 2.6%, from 40.5 GWh to 39.5 GWh, when the standby loss rate is assumed to be 15%; by 3.5%, from 40.5 GWh to 39.1 GWh, at an assumed standby loss rate of 20%; and by 4.4%, from 40.5 GWh to 38.7 GWh, when the standby loss rate is 25%. The generation cost also changes accordingly. Installing an ESS with an operating rate of 15% increases the generation cost by 13.3%, from 153.9 won/kWh to 174.3 won/kWh, when the standby loss rate is 15%; by 14.3%, from 153.9 won/kWh to 175.9 won/kWh, when the standby loss rate is 20%; and by 15.3%, from 153.9 won/kWh to 177.5 won/kWh, when the standby loss rate is 25%. The rate of increase in generation cost rises by 1.47 times compared with the rate of increase in investment costs when the loss rate is assumed to be 15% and the ESS operating rate to be 15%, and by 1.70 times when the loss rate is 25% and the ESS operating rate is 15%.

When a total charging and discharging and standby loss rate of 15% is assumed, using ESS at 15% power compensation for solar power and wind power generators raises the cost per kWh (simple LCOE) by 29.3 won per kWh for solar power and by 20.4 won for wind power compared with a situation where ESS is not used as such. ESS is evidently a more economical technology alternative than reserve power sources such as gas turbines and coal power. The ESS operating rate (compared with the generation facility) in which the cost of power compensation becomes similar to that for gas turbines is about 56% compared with solar power and about 78% compared with wind power. The ESS operating rate in which the cost of power compensation becomes similar to that for coal power is about 31% compared with solar power and about 43% when compared with wind power. In other words, it is evident that when ESS is used for power compensation, it can be a more economical alternative at a higher operating rate when installed in wind power generators compared with solar power.

In the case of wind-power generators, the amount of generation used as output variation by ESS in wind power plus ESS (10%) generators is 4,052 MWh in the first year. Applying the difference between the normal price of coal power and the normal price of LNG and coal power to this amount of generation results in total benefits in the first year of about 238.4 million won and 412.9 million won, respectively. Reflecting the benefits of using methods such as solar power generation in the LCOE computation results in corresponding figures of 159.6 won/kWh and 154.0 won/kWh. The generation cost with wind power alone, the generation cost of wind power +ESS, and the LCOE, taking into consideration the system benefits of wind power +ESS. When only wind power is installed, the generation price is 153.9 won/kWh, and when ESS is added to this, the generation price rises by 13.5 won/kWh, to 167.4 won/kWh. When the system benefits are taken into account, the generation cost becomes 154.0 won/kWh, which is 13.4 won/kWh lower than the generation costs of wind power plus ESS—almost at the same level as that of the generation cost for wind power alone.

APPENDIX C

C-1. MODELING AND SIMULATION TOOLS FOR ANALYSIS OF BATTERY ENERGY STORAGE SYSTEM PROJECTS

Although model-supported analysis of stationary storage systems has been conducted for specific storage technologies and applications, little existing work summarizes well the present modeling and simulation methods. Several distinct objectives often require specific approaches.

A noncomprehensive overview of some readily available modeling tools, with their respective aims, strengths, and shortcomings, is presented below.

Table C.1: Available Modeling Tools

Modeling Tool	Description
PerModAC	PerModAC was developed by researchers at the University of Applied Sciences in Berlin (HTW Berlin), Germany, to model the efficiency of PV–BESSs. While the model features an integrated approach, including all components relevant to efficiency-modeling of PV–BESSs (battery, inverter, standby, and energy management system control), the tool, in its present version, is confined to AC coupling of BESSs and does not allow modeling of battery aging.
Storage Value Estimation Tool (StorageVET)	StorageVET, developed by the US Electric Power Research Institute (EPRI), is focused on profitability analysis of utility-scale storage projects. This web-based tool allows straightforward value estimation for BESSs based on a set of predefined technical and economic input parameters.
Battery Lifetime Analysis and Simulation Tool (BLAST)	BLAST is a toolkit provided by NREL in the US. The focus is on LIB modeling, based on detailed battery aging degradation analysis and functionality, for long-term battery lifetime prediction. For use in automotive and stationary applications, several specialized variants exist (e.g., BLAST for Behind-the-Meter Applications, for peak-shaving application; BLAST for Stationary Applications, for other utility-scale applications).
Simulation of Stationary Energy Storage Systems (SimSES)	SimSES is a modular object-oriented tool chain initiated and coordinated by researchers at the Technical University of Munich in Germany. The MATLAB code–based tool-kit allows holistic technical and economic modeling of storage systems in variable applications. It features pre-implemented battery models (e.g., variants of LIB cells), and includes storage system and grid integration components (e.g., thermal management, power electronics components) and some exemplary use cases (e.g., PV–BESS, utility-scale control reserve).
Hybrid Optimization of Multiple Energy Resources (HOMER)	HOMER, originally developed by NREL in the US, is available as a commercial tool for microgrid modeling and optimization featuring fossil generation, renewable power sources, storage, and load management. The focus is therefore not on the battery itself but on the full grid level. For microgrid applications, straightforward cost optimization can be done with the help of HOMER, and storage system modeling features estimation of efficiency, self-discharge, and aging.
System Advisor Model (SAM)	SAM is a tool specifically designed for lead–acid and LIB modeling in PV–BESS applications. The battery modeling is based on detailed capacity, voltage, and thermal and lifetime sub-models, which can be parameterized, e.g., with the use of battery data sheets. For time-series calculations, user-defined operation strategies are required. In conjunction with financial modeling parameters, SAM allows straightforward techno-economic analysis. Like PerModAC, in its present version, SAM is limited to AC coupling of storage.
Modeling and Optimization of DERs-Based Systems (MODERS)	MODERS is a tool package aimed at the optimal sizing and economic evaluation of distributed energy resources in various applications. The package consists of a number of business model design programs, each of which was designed for a specific application such as load management, renewable-generation sale, energy prosumer, or community energy service. MODERS leads nontechnical users to readily follow preestablished procedures, offering a series of functions for assessing available energy from DERs, establishing optimal scheduling, estimating operating benefits, and finally determining optimal capacity. The tool package covers various DERs but is highly focused on BESS in terms of optimal sizing.

AC = alternating current, BESS = battery energy storage system, DER = distributed energy resource, LIB = lithium-ion battery, MATLAB = matrix laboratory, NREL = National Renewable Energy Laboratories, PbA = lead–acid, PV = photovoltaic, US = United States.

Source: Korea Battery Industry Association 2017 "Energy storage system technology and business model".

APPENDIX D
D-1. BATTERY ENERGY STORAGE SYSTEM IMPLEMENTATION EXAMPLE

Frequency Regulation at Sokcho Substation, Republic of Korea

Table D.1: Sokcho BESS Equipment Specifications

Component	Specifications	Quantity
PCS	Containers (4 MW each)	6
PMS	Human machine interface (HMI)	
Transformers	22.9 kV/440 V, 5.2 MVA	6
Cabling	AC/DC cable	

AC = alternating current, BESS = battery energy storage system, DC = direct current, MVA = megavolt-ampere, MW = megawatt, PCS = power conversion system, PMS = power management system, V = volt.

Source: Korea Battery Industry Association 2017 "Energy storage system technology and business model".

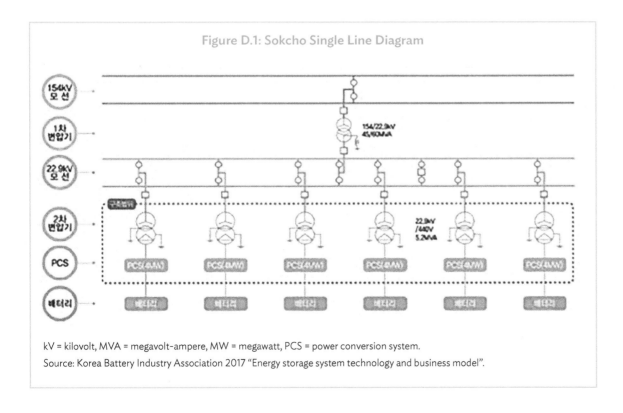

Figure D.1: Sokcho Single Line Diagram

kV = kilovolt, MVA = megavolt-ampere, MW = megawatt, PCS = power conversion system.

Source: Korea Battery Industry Association 2017 "Energy storage system technology and business model".

Figure D.2: Sokcho Site Plan

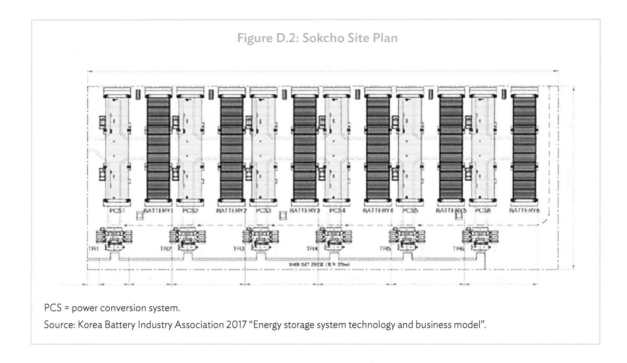

PCS = power conversion system.
Source: Korea Battery Industry Association 2017 "Energy storage system technology and business model".

Figure D.3: Bird's-Eye View of Sokcho Battery Energy Storage System

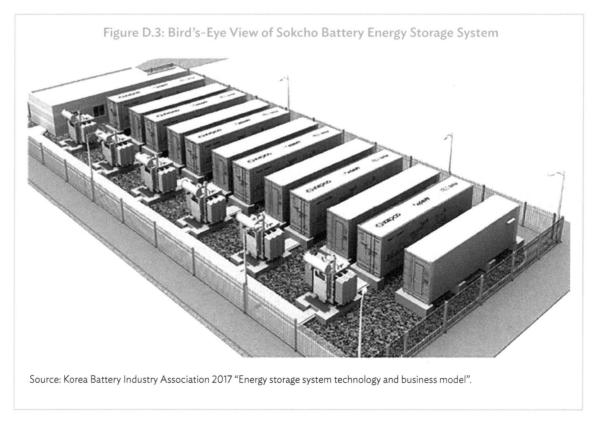

Source: Korea Battery Industry Association 2017 "Energy storage system technology and business model".

Figure D.4: Sokcho Battery Energy Storage System

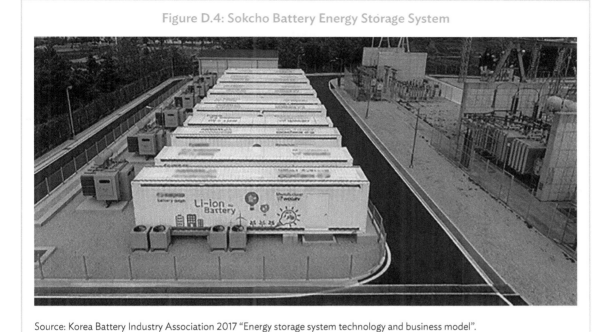

Source: Korea Battery Industry Association 2017 "Energy storage system technology and business model".

Table D.2: Renewable Integration

1.5 MW/3.2 MWh
Gasa Island, Republic of Korea

KEPCO/Renewable (PV, wind) Integrated

Wind: 98 MW
ESS PCS: 32 MW

Elkins, West Virginia, United States

continued on next page

Table D2 *continued*

| Solar PV: 28 MW |
| ESS PCS: 17 MW |
| |
| Chitose, Hokkaido, Japan |

ESS = energy storage system, KEPCO = Korea Electric Power Corporation, MW = megawatt, MWh = megawatt-hour, PCS = power conversion system, PV = photovoltaic.

Source: Korea Battery Industry Association 2017 "Energy storage system technology and business model".

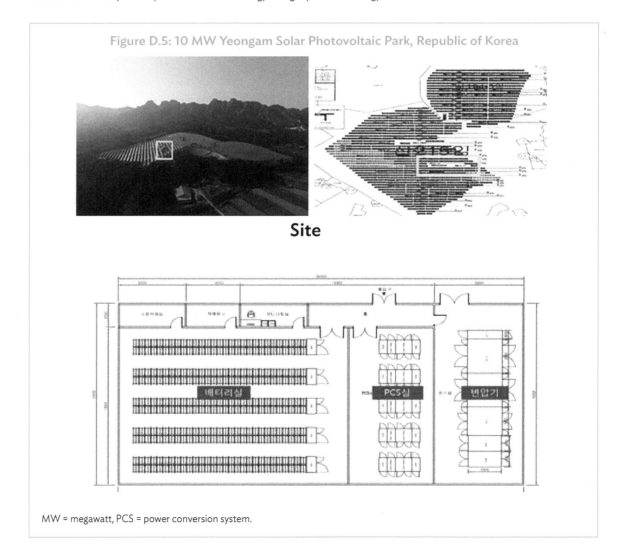

Figure D.5: 10 MW Yeongam Solar Photovoltaic Park, Republic of Korea

Site

MW = megawatt, PCS = power conversion system.

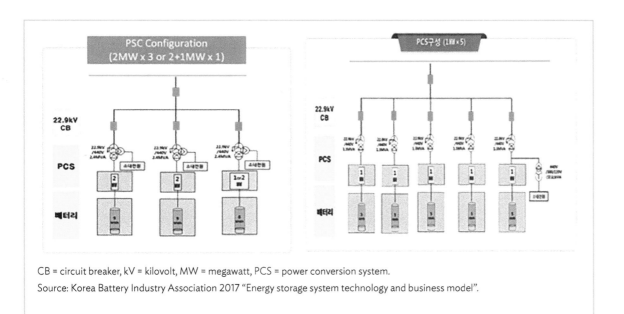

CB = circuit breaker, kV = kilovolt, MW = megawatt, PCS = power conversion system.

Source: Korea Battery Industry Association 2017 "Energy storage system technology and business model".

Configuration	Wind + ESS	PV + ESS	PV + Wind + ESS
Site	Chonnam National University	Kolon Global, Sebang	Sebang, Mokpo National University, JIAT, KETI
Specifications	100 kW (wind) + 100 kW (bidirectional converter) + 56.6 kWh ESS	15 kW (PV) + PV PCS + 52 kWh ESS	10 kW (PV) + 1 kW (wind) + 60 kW (bidirectional converter) PCS + ESS (200 kWh) + 50 kW EV (high-speed battery charger)
ESS capacity	56.6 kWh (Ni–MH, 80 Ah, 590 cell)	52 kWh (Ni–MH, 100 Ah, 440 cell)	Rapid: 200 kWh (Ni–MH, 300 Ah, 550 cell) Full: 12 kWh (Ni–MH, 50 Ah, 210 cell)

Ah = ampere-hour, ESS = energy storage system, EV = electric vehicle, JIAT = Jeonbuk Institute of Automotive Technology, KETI = Korea Electronics Technology Institute, kW = kilowatt, kWh = kilowatt-hour, Ni–MH = nickel–metal hydride, PCS = power conversion system, PV = photovoltaic.

Source: Korea Battery Industry Association 2017 "Energy storage system technology and business model".

Figure D.6: Peak Shaving at Douzone Office Building, Republic of Korea

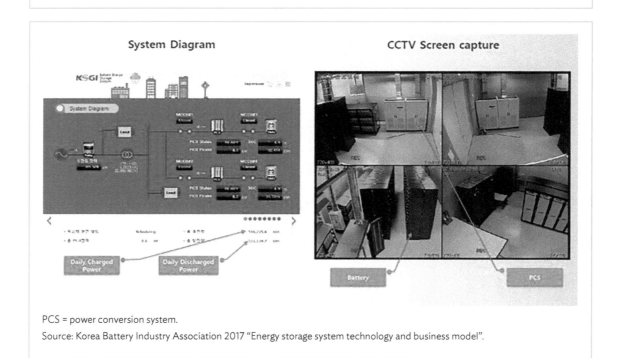

ESS = Energy Storage System, kW = kilowatt, kWh = kilowatt-hour.
Source: Korea Battery Industry Association 2017 "Energy storage system technology and business model".

PCS = power conversion system.
Source: Korea Battery Industry Association 2017 "Energy storage system technology and business model".

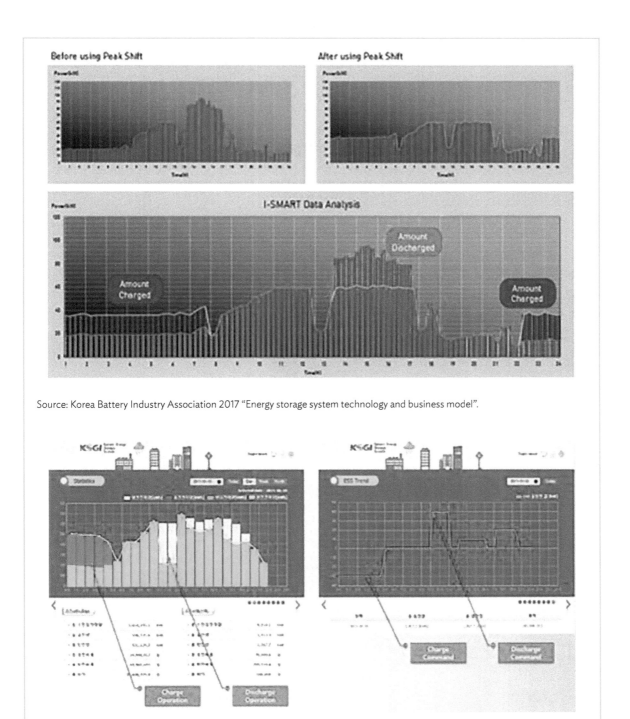

Source: Korea Battery Industry Association 2017 "Energy storage system technology and business model".

Source: Korea Battery Industry Association 2017 "Energy storage system technology and business model".

Table D.3: Black Start Capability

	Smart Electric Power Association (SEPA), Southern California, US 33 MW/20 MWh (BESS) + 44 MW (combined-cycle gas turbine)
	Vulkan Energiewirtschaft Oderbrücke (Vulkan Energy Bridge) GmbH, Germany 720 kWh (BESS) + 400 kW (gas turbine)

BESS = battery energy storage system, kW = kilowatt, kWh = kilowatt=hour, MW = megawatt, MWh = megawatt-hour.

Source: Korea Battery Industry Association 2017 "Energy storage system technology and business model".

Figure D.7: First Microgrid System on Gapa Island

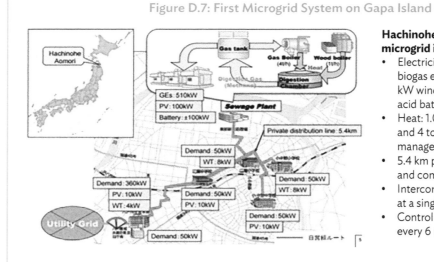

Hachinohe: First stage of microgrid in Japan (2003–2007)
- Electricity: 510 kW (170 X 3) biogas engines, 130 kW PV, 20 kW wind turbines, 100 kW lead–acid battery
- Heat: 1.0 ton/h wood boiler and 4 tons/h gas boiler Energy management
- 5.4 km private lines (electricity and communication)
- Interconnected with utility grid at a single point
- Control error target: within 3% every 6 minutes moving average

Source: "Microgrid Projects," [Online]. Available: http://microgridprojects.com/microgrid/hachinohe-microgrid/.

continued on next page

Figure D.8: Sendai Microgrid project (Shimizu 2013)

Sendai Microgrid Project

- Constructed as a 4-year demonstration project
- Entrusted by NEDO
- Technical feature: Multiple Power Quality Microgrid (MPQM)
  - Desired power quality varies by customer in the microgrid
  - MPQM enables power supply with different levels of power quality for different customers in the area

Source: Y. Shimizu, "Microgrid related activities in Japan," in New Energy and Industrial Technology Development Organization (NEDO), Sep.11, 2013.

genset = generator set, kW = kilowatt, km = kilometer, NEDO = New Energy and Industrial Technology Development Organization, PV = photovoltaic.

Source: Korea Battery Industry Association 2017 "Energy storage system technology and business model".

Figure D.9: System Configuration of Microgrid on Gapa Island

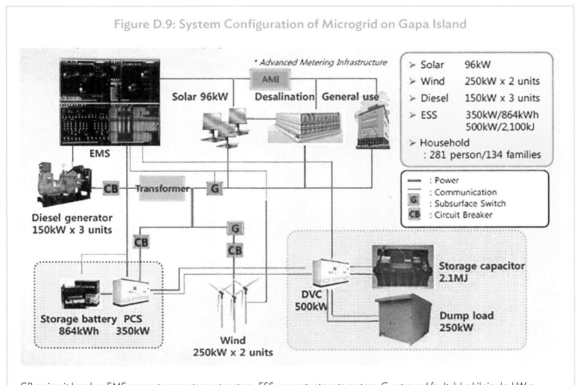

CB = circuit breaker, EMS = energy management system, ESS = energy storage system, G = ground fault, kJ = kilojoule, kW = kilowatt, kWh = kilowatt-hour, MJ = megajoule, PCS = power conversion system.

Source: Korea Battery Industry Association 2017 "Energy storage system technology and business model".

APPENDIX E
BATTERY CHEMISTRY

Lead–Acid (PbA) Battery

This cell is a widely applied type of secondary cell, used extensively in vehicles and other applications requiring high values of load current. Its primary benefits are low capital costs, maturity of technology, and efficient recycling.

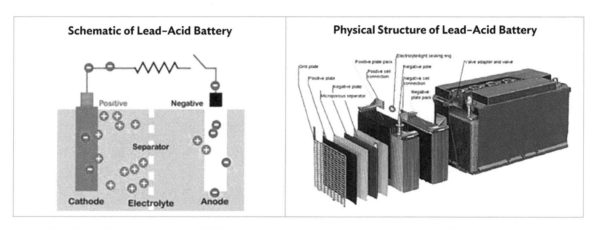

Source: Korea Battery Industry Association 2017 "Energy storage system technology and business model".

Nickel–Cadmium (Ni–Cd) Battery

A nickel–cadmium (Ni–Cd) battery is a rechargeable battery used for portable computers, drills, camcorders, and other small battery-operated devices requiring an even power discharge.

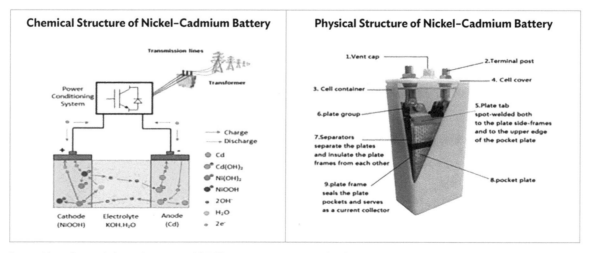

Source: Korea Battery Industry Association 2017 "Energy storage system technology and business model".

Nickel–Metal Hybrid (Ni-MH) Battery

The nickel–metal hydride battery chemistry is a hybrid of the proven positive electrode chemistry of the sealed nickel–cadmium battery with the energy storage features of metal alloys developed for advanced hydrogen energy storage concepts (Moltech Power Systems 2018[1])

Ni–MH batteries outperform other rechargeable batteries and have higher capacity and less voltage depression.

Ni–MH batteries are currently finding widespread application in high-end portable electronic products, where battery performance parameters, notably run time, are a major consideration in the purchase decision.

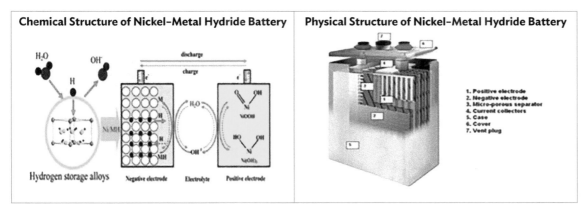

Source: Korea Battery Industry Association 2017 "Energy storage system technology and business model".

Lithium-Ion Battery

Lithium-ion batteries have the highest energy density and are safe. No memory or scheduled cycling is required to prolong battery life. Lithium-ion batteries are used in electronic devices such as cameras, calculators, laptop computers, and mobile phones, and are increasingly being used for electric mobility.

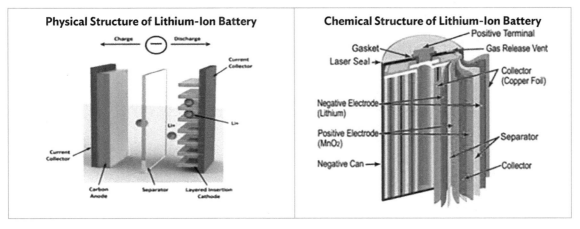

Source: Korea Battery Industry Association 2017 "Energy storage system technology and business model".

---

[1]   Source: M. P. systems, "NiMH Technology," 2018. [Online]. Available: https://www.tayloredge.com/.../Batteries/Ni-MH_Generic.pdf.

Sodium–Sulfur (Na–S) Battery

The sodium–sulfur battery, a liquid-metal battery, is a type of molten metal battery constructed from sodium (Na) and sulfur (S). It exhibits high energy density, high efficiency of charge and discharge (89%–92%), and a long cycle life, and is fabricated from inexpensive materials.

However, because of its high operating temperatures of 300°C–350°C and the highly corrosive nature of sodium polysulfides, such cells are primarily used for large-scale nonmobile applications such as electricity grid energy storage.[2]

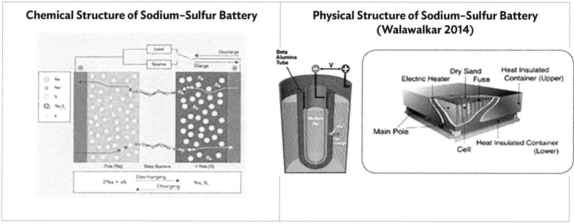

Source: D. R. Walawalkar, "Li-ion Battery Technology - IRENA," [Online]. Available:www.irena.org/.../IRENA_walawalkar_status_storage_final.pptx.

---

2   Advanced Thin Film Sodium Battery. China Energy Storage, en.escn.com.cn/Tools/download.ashx?id=131.

Redox Flow Battery (RFB)

RFBs are charged and discharged by means of the oxidation–reduction reaction of ions of vanadium or the like. They have excellent characteristics: a long service life with almost no degradation of electrodes and electrolytes, high safety because of their being free of combustible materials, and availability for operation under normal temperatures.

Schematic of Redox Flow Battery (Walawalkar 2014)	**Physical Structure of Redox Flow Battery (Fischer and Tuebke 2018)**
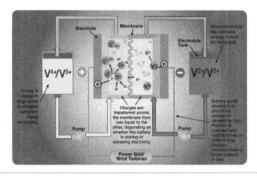	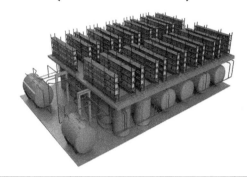
Source: D. R. Walawalkar, "Li-ion Battery Technology - IRENA," [Online]. Available: www.irena.org/.../IRENA_walawalkar_status_storage_final.pptx.	**Source:** P. D. J. T. Dr. P. Fischer, "Redox Flow Batteries for stationary storage," 29 1 2018. [Online]. Available: https://www.steag.in/sites/default/files/Paper%2011%20-%20Redox%20Flow%20Batteries%20for%20stationary%20storage%20applications.ppt%20%5BCompatibility%20Mode%5D.pdf.

COMPARISON OF TECHNICAL CHARACTERISTICS OF ENERGY STORAGE SYSTEM APPLICATIONS

ESS Application	Required Response Time	Reference Duty Cycle	Energy Discharge Cycle Duty	ESS Unit Power (MW$_{AC}$)	ESS AC Voltage (kV)	Full Power Discharge Duration	Basis for Economic Benefits
3-hour load shift	10 minutes	Scheduled 3-hour discharge	60 d/y, 1 event/d	1–200	4.2–115	3 hours	Market rates
10-hour load shift	10 minutes	Scheduled 10-hour discharge	250 d/y, 1 event/d	1–200	4.2–115	10 hours	Market rates
Renewables time shift	1 minute	Optimized by tech	According to reference wind profile	2–200	4.2–34.5	5–12 hours (except CAES; varies)	Various*
Renewables forecast hedging	20 milliseconds	Optimized by tech	According to reference wind profile	2–200	4.2–34.5	5–12 hours (except CAES; varies)	Various*
Fluctuation suppression	20 milliseconds	Continuous cycling	90 cycles/hr	2–50	4.2–34.5	10 seconds	Various*
Short-duration power quality	20 milliseconds	Hot standby	100 events/y, 5 events/d, 1 event/hr	1–50	4.2–34.5	5 seconds	Cost of alternative solutions
Long-duration power quality	20 milliseconds	Hot standby for infrequent events	1 event/y	1–50	4.2–34.5	4 hours	Cost of alternative solutions
Frequency excursion suppression	20 milliseconds	Hot standby	10 events/y, 1 event/d	10–500	4.2–750	15 minutes	Cost of alternative solutions
Grid frequency support	20 milliseconds		24 events/y, 1 event/d	2–200	4.2–34.5	10–30 minutes	Various*
Angular stability	20 milliseconds	Hot standby	10 events/y, 1 event/d, 20 cycles/event	10–500	4.2–750	1 second	Cost of alternative solutions
Voltage stability	20 milliseconds	Hot standby	10 events/y, 1 event/d	10–500	4.2–750	1 second	Cost of alternative solutions
Transmission curtailment	1 min	Optimized by tech	According to reference wind profile	2–200	4.2–34.5	5–12 hours (except CAES; varies)	Various*

AC = alternating current; CAES = compressed-air energy storage; d = day; d/y = days per year; ESS = energy storage system; hr = hour; kV = kilovolt; min = minute; MWAC = megawatt, alternating current; y = year.

* Various bases for economic benefits of these applications, including capitalized costs and benefits of alternative systems, market rates, tax credits, and green price premiums.

Source: Carnegie et al. (2013).

SUMMARY OF GRID STORAGE TECHNOLOGY COMPARISON METRICS

Metric	Hydro	Flywheel	Lead–Acid	Ni–MH	Thermal	Li	Flow	Liquid Metal	Compressed Air
Specific energy (kW/kg)	0.3–1.33	5–200	30–50	30–90	10–250	90–250	10–90	100–240	3.2–60
Energy density (kWh/vol)	0.5–1.33	0.25–424	25–90	38.9–300	25–370	94–500	5.17–70	150–345	0.4–20
Specific power (W/kg)	0.001–0.12	400–30,000	25–415	50–1,000	10–30	8–2,000	5.5–166	14.29–260	2.2–24
Cycle life	20–50k	Indefinite	200–2k	300–10k	Indefinite	500–10k	10k+	5k–10k+	5k–20k+
Useful life	50–60	20	10–15	5–10	20+	5–15	5–20	10	25–40
Life cycle	Near universal life with maintenance	Near universal life with maintenance	Useful life varies by depth of discharge and application, variations by chemistry	Allows deeper discharge and more stable storage, variations by chemistry	Thermal salts not yet proven; passive storage varies by technology	Useful life varies by depth of discharge and other applications, variations by chemistry	Moving parts require intermittent replacement	Not yet proven	Near universal life with maintenance
Cost per kWh	$1–$291	$200–$150,000	$50–$1,100	$100–$1,000	$1–$137	$200–$4,000	$100–$2,000	$150–$900	$1–$140
Environmental impact	High/Mixed	Low	High	High/Medium	Low	High/Medium	Medium	Low	Low/Medium
Pros	Large power capacity, positive externalities	Very fast response, high specific power, low cost, long life	Mature technology with established value proposition	Deep discharge capacity, reliable, high energy density	Could pair with waste heat generation, scalable, low cost, large scale	Flexible uses, very fast response and high specific power	Large storage capacity, cheap materials	High capacity, fast response, cheap materials, highly stable, temperature tolerant	Low cost, large scale, mature technology paired with gas turbines
Cons	Geographically limited, expensive construction, low energy density and environmentally damaging	Low energy density	Low life cycle, toxic materials, flammability risk	Some toxic variations, less specific power than Li, high self-discharge, high memory effect	Not fully commercialized or not electrified	Safety concerns, low depth of discharge, corrosion, self-discharge, and efficiency loss over time	Space requirements, economic efficiency in multiple applications	Untested in commercial use, persistent technology issues	Geographically limited, not scalable

k = thousand, kg = kilogram, kWh = kilowatt-hour, Li = lithium, Ni–MH = nickel–metal hydride, W = watt.
Source: Hart, Bonvillian, and Austin (2018).

REFERENCES

Battery University. 2018a. BU-201: How Does the Lead Acid Battery Work? 31 May. http://batteryuniversity.com/learn/article/lead_based_batteries.

———. 2018b. BU-203: Nickel-based Batteries. 31 May. http://batteryuniversity.com/learn/article/nickel_based_batteries.

———. 2018c. BU-205: Types of Lithium-ion Batteries. 31 May. http://batteryuniversity.com/learn/article/types_of_lithium_ion.

Carnegie, Rachel, Douglas Gotham, David Nderitu, and Paul V. Preckel. 2013. Utility Scale Energy Storage Systems: Benefits, Applications, and Technologies. State Utility Forecasting Group, Purdue University, Indiana, US. June.

China News Service. n.d. Advanced Thin Film Sodium Sulfur Battery. en.escn.com.cn/Tools/download.ashx?id=131.

Confais, Eric, and Ward van den Berg. 2017. Business Models in Energy Storage: Energy Storage Can Bring Utilities Back into the Game. Roland Berger Focus. May.

Doughty, D. H., and E. Peter Roth. 2012. A General Discussion of Li Ion Battery Safety. Electrochemical Society Interface 21 (2): 37–44. DOI: 10.1149/2.F03122if.

Electric Power Research Institute (EPRI). 2010. Electricity Energy Storage Technology Options: A White Paper Primer on Applications, Costs, and Benefits. Palo Alto, California, US. http://large.stanford.edu/courses/2012/ph240/doshay1/docs/EPRI.pdf

Enel Green Power. 2016. Integrating Renewable Power Plants with Energy Storage. 7 June. http://www.iefe.unibocconi.it/wps/wcm/connect/29b685e1-8c34-4942-8da3-6ab5e701792b/Slides+Lanuzza+7+giugno+2016.pdf?MOD=AJPERES&CVID=lle7w78.

Energy and Environmental Economics, Inc. (E3). 2011. WECC Efficient Dispatch Toolkit (EDT) Phase 2 Energy Imbalance Market (EIM) Benefits Analysis & Results (October 2011 Revision). Prepared for the Western Electricity Coordinating Council, Utah, US.

Fischer, P., and J. Tuebke. 2018. Redox Flow Batteries for Stationary Storage Applications. 29 January. https://www.steag.in/sites/default/files/Paper%2011%20-%20Redox%20Flow%20Batteries%20for%20stationary%20storage%20applications.ppt%20%5BCompatibility%20Mode%5D.pdf.

Hart, David M., William Bonvillian, and Nathaniel Austin. 2018. Energy Storage for the Grid: Policy Options for Sustaining Innovation. *MIT Energy Initiative Working Paper Series*. Massachusetts Institute of Technology, US. April.

Hesse, Holger C., Michael Schimpe, Daniel Kucevic, and Andreas Jossen. 2017. Lithium-Ion Battery Storage for the Grid: A Review of Stationary Battery Storage System Design Tailored for Applications in Modern Power Grids. *Energies* 10 (12). https://doi.org/10.3390/en10122107.

International Renewable Energy Agency (IRENA). 2015. *Battery Storage for Renewables: Market Status and Technology Outlook.* Abu Dhabi.

———. 2017. *Electricity Storage and Renewables: Costs and Markets to 2030.* Abu Dhabi.

Jim. 2014. Used Nissan EV Batteries Now Provide Grid Scale Storage (blog). *Vehicle to Grid UK* 11 May. http://www.v2g.co.uk/2014/05/used-nissan-ev-batteries-now-provide-grid-scale-storage/.

Johnson, Jay, Benjamin Schenkman, Abraham Ellis, Jimmy Quiroz, and Carl Lenox. 2011. *Initial Operating Experience of the La Ola 1.2-MW Photovoltaic System.* Sandia National Laboratories Report SAND2011-8848.

Kane, Mark. 2015. Bosch Cooperates With BMW And Vattenfall In Second Life Battery Project. *Inside EVs* 9 February. https://insideevs.com/bosch-cooperates-with-bmw-and-vattenfall-in-second-life-battery-project/.

———. 2017. Renault To Enter Home Battery Market With Repurposed EV Batteries. *Inside EVs* 26 June. https://insideevs.com/renault-repurposed-ev-batteries-ess/.

McKenna, Eoghan, John Barton, and Murray Thomson. 2017. Short-Run Impact of Electricity Storage on CO2 Emissions in Power Systems with High Penetrations of Wind Power: A Case-Study of Ireland. *Journal of Power and Energy* 231 (6): 590–603. https://doi.org/10.1177/0957650916671432.

Microgrid Media. n.d. Microgrid Projects: Hachinohe Microgrid, Japan. Minnesota. http://microgridprojects.com/microgrid/hachinohe-microgrid/.

Moltech Power Systems. 2018. NiMH Technology: A Generic Overview. Florida. https://www.tayloredge.com/reference/Batteries/Ni-MH_Generic.pdf.

Morris, Charles. 2015. Nissan, GM and Toyota Repurpose Used EV Batteries for Stationary Storage. *Charged Electric Vehicles Magazine* 17 June. https://chargedevs.com/newswire/nissan-gm-and-toyota-repurpose-used-ev-batteries-for-stationary-storage/.

Reid, Gerard, and Javier Julve. 2016. *Second-Life Batteries As Flexible Storage For Renewable Energy.* Report prepared for the German Renewable Energy Federation (BEE) and the Hanover Trade Fair. April.

Sandia National Laboratories. 2013. DOE/EPRI 2013 Electricity Storage Handbook. Prepared in collaboration with the National Rural Electric Cooperative Association (NRECA) for the US Department of Energy and the Electric Power Research Institute, US. New Mexico. https://prod.sandia.gov/techlib-noauth/access-control.cgi/2015/151002.pdf.

Shimizu, Yasuhiro. 2013. Micro-grid Related Activities in Japan. New Energy and Industrial Technology Development Organization (NEDO). 11 September.

Sumitomo Electric Industries Ltd.. n.d. Redox Flow Battery. Osaka, Japan. http://global-sei.com/products/redox/.

Thorbergsson, Egill, Vaclav Knap, Maciej Swierczynski, Daniel Stroe, and Remus Teodorescu. 2013. Primary Frequency Regulation with Li-Ion Battery Based Energy Storage System: Evaluation and Comparison of Different Control Strategies. *Intelec 2013*. 35th International Telecommunications Energy Conference. Hamburg, Germany. 13–17 October.

University of Minnesota Energy Transition Lab, Strategen Consulting, and Vibrant Clean Energy. 2017. *Modernizing Minnesota's Grid : An Economic Analysis of Energy Storage Opportunities*. Report prepared for the Minnesota Energy Storage Strategy Workshop. 11 July.

Walawalkar, Rahul. 2014. Energy Storage Technology Overview. Presented at the International Electricity Storage Policy and Regulation Workshop of the International Renewable Energy Agency (IRENA). New Delhi, India. 3 December.